# SCIENTIFIC EXCELLENCE AND EQUALITY AT UNIVERSITY



Based on the EU-funded project RESET (Redesigning Equality and Scientific Excellence Together) and written by members of seven European partner universities representing different disciplines, this book combines theory and practice to discuss and reflect on the concept of scientific excellence within higher education. It considers the pitfalls and the need to revisit an understanding of inclusivity by asking how scientific excellence and professional equality can be made compatible in practice.

Split into three key sections, the first part of the book is devoted to the question of scientific excellence in relation to existing gender inequalities within universities. It includes a discussion on the objectivity of scientific excellence based on feminist literature devoted to the reproduction of inequalities in universities and data collected during the RESET project. The second part is dedicated to the actions taken to achieve more inclusive excellence and the third part finishes by focussing on training on gender and diversity issues and the resistance that such workshops generate.

The book offers a comprehensive guide to fostering an academic culture that aligns scientific excellence with equality, diversity, and inclusion, providing researchers and decision-makers alike with practical tools for achieving inclusive excellence within European universities, while at the same time knowing how to deal with difficulties and resistance.

**Marion Paoletti** is Professor of Political Science at the University of Bordeaux, France and Scientific Director of the RESET project.

**Marisa Matias** is Professor of Psychology at the University of Porto, Portugal and Scientific Coordinator of RESET at the University of Porto.

**Aleksandra Różalska** is Assistant Professor and Head of the Women Studies Centre, University of Lodz, Poland.

# SCIENTIFIC EXCELLENCE AND EQUALITY AT UNIVERSITY

## Experiencing the RESET Project

*Edited by Marion Paoletti, Marisa Matias, and Aleksandra Różalska*

Routledge
Taylor & Francis Group
LONDON AND NEW YORK

Designed cover image: © Getty Images

First published 2026
by Routledge
4 Park Square, Milton Park, Abingdon, Oxon OX14 4RN

and by Routledge
605 Third Avenue, New York, NY 10158

*Routledge is an imprint of the Taylor & Francis Group, an informa business*



The Open Access conversion of this book was funded from the European Union's Horizons 2020 Framework Program for Research and Innovation under Grant Agreement n° 101006560

European Commission



*British Library Cataloguing-in-Publication Data*
A catalogue record for this book is available from the British Library

ISBN: 978-1-032-91650-7 (hbk)
ISBN: 978-1-032-91648-4 (pbk)
ISBN: 978-1-003-56434-8 (ebk)

DOI: 10.4324/9781003564348

Typeset in Times New Roman
by Apex CoVantage, LLC

# CONTENTS

# FIGURES

# TABLES

# CONTRIBUTOR BIOS

**Areti Ampatzoglou** is an economist with an interdisciplinary academic background, holding an MSc degree in Information Systems and a PhD in Software Economics. Having many years of teaching experience, she is currently serving as a teaching assistant at the School of Informatics at the Aristotle University of Thessaloniki. In addition to her educational role, Areti has actively contributed to European research projects for the past six years. In her present role as a research associate at the Data and Web Science Lab (Datalab) of Aristotle University, she is a member of the research team for the RESET H2020 project. Her primary focus involves the collection and analysis of data to facilitate data-driven decision-making processes and the shaping of gender equality policies.

**Lamprini Argyropoulou** holds a BSc in Economics and an MSc degree in Journalism and Digital Media. She has over ten years of experience in marketing and project communication design. She has worked as an executive coordinator for the Thessaloniki Convention Bureau. She got involved in the MICE (meetings, incentives, conventions, and exhibitions) industry, representing Greek destinations in international exhibitions and B2B forums. She was a member of the "Liberty International Tourism Organization" and the "Professional Conference Management Association", working as a project and marketing manager. She has also had experience as a researcher and communication advisor at Aristotle University of Thessaloniki for EU-funded educational projects related to the growth of nautical and watersports tourism in Greece and ESG communication. Currently, she is a PhD candidate at the School of Tourism at Ionian University.

**Maria Isabel Dias** holds a PhD (2003) and a Habilitation (2009) in Sociology. She has been a researcher at the Institute of Sociology since 1990, coordinating the Family, Ageing and Gender research group. She is also a full professor at the Department of Sociology, Faculty of Arts and Humanities, University of Porto. Her research focuses on domestic violence and gender inequalities, women's work and occupational health, and elder abuse through her participation in various national and international research projects. She is also working on gender inequalities in academia and science and on implementing gender equality plans. She has published eight books, 30 chapters, and several academic articles in peer-reviewed journals.

**Eva Durall Gazulla,** an academy postdoctoral researcher at the INTERACT research unit at University of Oulu, Finland. Her expertise areas include media and technology, design and learning using critical approaches as well as participatory design and futures techniques to engage participants in research. Currently, she is involved as a research leader in various projects exploring democratic values in technology design and use with young people (co-PI in Critical ChangeLab and PI in Critical DataLit projects).

**Maxime Forest,** a political scientist, is a lecturer and researcher at Sciences Po (OFCE, PRESAGE). His work focuses on the comparative analysis of gender equality policies and their Europeanisation. He has coordinated and evaluated several European projects on gender in research. He is the author of the chapter on "Europeanisation" in Abels, G., Krizsán, A., MacRae H., and Van der Vleuten, A. (eds.) The Routledge Handbook of Gender and EU Politics (2021) and co-authored with Maria Bustelo and Lucy Ferguson The Politics of Feminist Knowledge Transfer, Palgrave MacMillan (2016) and The Europeanisation of Gender Equality Policies, Palgrave MacMillan (2012), with Emmanuela Lombardo.

**Dimitra Hadjipavlou-Litina** is a full professor in Medicinal Chemistry. From 2021 to 2024, she is the Principal Investigator for AUTh University for the HORIZON research project *Redesigning Equality and Scientific Excellence Together* (RESET). She holds a number of positions as: member of national and international scientific societies, member of the board of the Hellenic National Academic Recognition and Information Center, participation in more than 500 national and international congresses, evaluator for national grants/international grants, chairman and vice chairman of the School of Pharmacy, member of the senate, and director of the postgraduate programme.

**Mateusz Hauk** holds a PhD in Social Sciences (Psychology) and is an assistant professor at the Institute of Psychology at the University of Lodz. Beyond research and teaching, he has worked as a trainer since 2005, delivering training sessions and development programmes on managerial and leadership skills, workplace relationships, communication and assertiveness, team building, facilitation and moderation, personal effectiveness (psychological and social capital), and implementing and managing change. When carrying out training sessions and workshops, he promotes the idea of "building satisfying relationships", emphasising respect and equality.

**Mervi Heikkinen**, PhD, Title of Docent, investigates feminist knowledge, intersectional gender equality, epistemic contribution, and sociomaterial entanglements in the context of knowledge production. Heikkinen has published over 50 academic works with a gender perspective. Heikkinen is a university lecturer in university pedagogy at the University of Oulu. Over the past three decades, she has contributed to Gender Studies through various Nordic and EC-funded projects. She leads the University of Arctic thematic network on Gender in Arctic knowledge production. In the RESET project, she led WP7, focusing on mainstreaming gender dimension in the design of research activities and products.

**Netta Iivari,** Professor of Information Systems, Research Unit Leader of Interact, University of Oulu. Her research aims at understanding and strengthening people's participation in shaping and making their digital futures. She has a strong background in co-design and participatory design as well as in the development and exploration of speculative, value-driven, critical, inclusive, and empowering design approaches. Her research includes studies on gender equality, addressing the contexts of higher education institutions and home energy technology adoption and use with a focus on associated values, practices, and discourses. She has acted as a PI in numerous Research Council of Finland and EU-funded research projects addressing these topics.

**Ninon Junca** is a graduate of Sciences Po Bordeaux and was a RESET project manager at the University of Bordeaux. She led the production of the joint declaration by university presidents on inclusive excellence and the production of relevant indicators to support this redefinition.

**Eleni Karachaliou** holds a MEng in Urban/Spatial Planning and Development Engineering from the Engineering Faculty of Aristotle University of Thessaloniki (AUTh) and a Master's degree in Geoinformatics, also from AUTh, as a fellow of the Onassis Foundation (2014–2015). Currently, she is a PhD candidate at the School of Spatial Planning and Development in AUTh. She has more than nine years of experience in national and international research, development, and innovation projects in several sectors as a research associate, consultant, and project manager. Since 2021, she has been a core team member in the implementation team of RESET H2020 project. She has also been part of the team designing, developing, and implementing the 1st Gender Equality Plan of the Aristotle University of Thessaloniki.

**Stelios Karamanidis** holds a BSc in Physics and an MSc degree in Nanosciences and Nanotechnologies from the Aristotle University of Thessaloniki (AUTh). He has more than a decade of experience as a research associate in various national and international projects. With a unique blend of expertise in research activities and technical skills in web development, he has actively contributed as a software engineer, to the creation and enhancement of multiple web applications in the realm of human-centred data science. Presently affiliated with AUTh's Data and Web Science Lab (Datalab), he has been part of the AUTh's research team of RESET H2020 project, mainly involved in and focused on the design, development, and evaluation of the data, dissemination, and web tools designing and developing tasks.

**Tuija Lämsä,** D.Sc. (Econ. & Bus. Adm.), Director of Bachelor's Programme, Oulu Business School, University of Oulu. Her main research interests are organisational behaviour and creativity in the era of change, focusing on leadership, knowledge management, and organisational learning (incl. Communities of Practice approach). Her research also includes studies on gender equality and scientific excellence in the context of higher education institutions. She is strongly research-orientated, and actively participates in research collaborations and projects, both in the international and regional scientific community. She also has a strong social impressiveness at many academic and administrative levels and communities, including not only the university field but also other stakeholders, as well as business environments.

**Alexandra Lopes**, Assistant Professor at the Department of Sociology of the University of Porto and a researcher at CETAPS. Her research focuses on social phenomena related to ageing, with a focus on LTC, formal and informal carers, family dynamics of care, and inequalities in care. She has coordinated and collaborated in several research projects about these topics (Horizon 2020, Horizon Europe, and FCT) and published over 90 publications. She is currently the director of the Iberian PhD programme in Social Sciences and Ageing.

**Sara Isabel Magalhães** has a degree and PhD in Educational Psychology, both from the University of Minho. She is a PhD researcher at the Psychology Centre of the University of Porto. She specialises in Gender Equality and is a member of the CIG's specialist list. She has collaborated and co-coordinated several funded research projects (European Commission Horizon 2020 SwafS; EEA GRANTS/FSE/CIG; POPH/QREN/FSE/CIG; POISE/CIG; FCT). She is a feminist activist and collaborates with several national NGOs.

**Kalliopi Malerou** holds a BSc in Computer Science from Aristotle University in Thessaloniki. With a diverse range of roles, she has gained valuable experience, showcasing her proficiency in software engineering and data analysis.

Currently employed as a Software Developer, she specialises in web and mobile development, primarily focusing on the front-end aspects of applications. In addition to her professional achievements, she has made academic contributions by co-authoring a publication titled "DIFCURV: A Unified Framework for Diffusion Curve Fitting and Prediction in Online Social Networks", which was published in Array in 2021 and can be found under https://doi.org/10.1016/j.array.2021.100100. Regarding the RESET project, Kalliopi made essential contributions to the data life cycle, handling collection, cleaning, and analysis with precision. Her contributions extended further as she actively supported the development and ongoing maintenance of the RESET platform, ensuring its functionality and reliability.

**Marisa Matias** is Doctor in Psychology and Assistant Professor at the Faculty of Psychology and Educational Sciences at the University of Porto. Scientific coordinator of RESET at U.Porto, she has published three books, 12 book chapters, and more than 50 scientific papers in peer-reviewed journals. She researches in the area of gender, work-family articulation, and family relationships through several national and international research projects. Her most recent investments have been in understanding parenting and work-life articulation from a cultural and international perspective, alongside addressing the role of HEIs as agents of change towards gender equality.

**Maria Michali** is a research associate at Aristotle University of Thessaloniki, Data and Web Science Lab (Datalab) and a member of AUTh's research team in the RESET H2020 project. Her work in RESET is centred around the impact and contribution of RESET to EU strategies and agendas, and to actions related to the Gender Equality Plan of AUTh. She is also pursuing her PhD studies at the University of York, School for Business and Society, with her study addressing the effectiveness of organisational gender policies.

**Viktoria Niebel** is a social scientist and research associate at the Science Support Center of the University Duisburg Essen, where she is responsible for gender consulting, career counselling for female early career researchers, and the promotion of equal opportunities at the Graduate Center Plus. She was a local project manager in the EU-funded project RESET: *Redesigning Equality and Scientific Excellence Together* at Ruhr University Bochum (RUB; 2021–2024). From 2017 to 2021, she was a research associate at the Faculty of Social Sciences and acted as deputy of the Central Equal Opportunities Officer for academic staff at RUB. As a doctoral candidate, she researches discourses and practices of mindfulness in Western societies at the Chair of Social Theory and Social Psychology at RUB. Furthermore, her academic interests are inter- and transculturality and intercultural communication and competence, diversity, and gender equity.

**Conceição Nogueira** is a full professor at the Faculty of Psychology and Educational Sciences of the University of Porto, where she is the director of the Doctoral Programme in Human Sexuality. She works in the field of Gender and Sexuality Studies from a feminist perspective. She has coordinated several projects on different topics that converge around critical social intervention in favour of Gender, Diversity, and Sexuality issues. Since 2018, she has been the president of the non-profit organisation *Gentopia – Association for Gender Diversity and Equality*. She is the author of several books in Portuguese (Portugal and Brazil) and numerous (inter)national publications on Gender Studies, Feminisms, and Sexualities.

**Pat O'Connor** is Professor Emeritus of Sociology and Social Policy at University of Limerick (UL) and Visiting Full Professor at Geary Institute, University College Dublin. She has also held visiting professorships in London, Aveiro, Linkoping, Deakin, and Melbourne. A member of the *National Review on Gender Equality in Irish Higher Education Institutions* (2016), her current research interests revolve around gender inequality in higher education with a focus on leadership and management, organisational structures and culture, excellence, micropolitics, and gender-equality-related interventions and support. She was the first female full professor (1997) and Faculty Dean (2000) at UL and the first female professor of sociology in Ireland (1997). Her 120 publications include eight books, over 80 peer-reviewed articles, chapters, reports, and policy pieces.

**Marion Paoletti** is a professor of Political Science at the University of Bordeaux. She was the scientific coordinator of the RESET project. She currently heads the Montesquieu Research Institute (IRM). She recently coordinated a controversy in the French journal *Travail, genre et société* titled "L'excellence scientifique: piège ou opportunité pour les femmes?" (with Sophie Pochic and Fanny Gallot, 2024). She co-edited the book *Politiques de la fusion* (LGDJ, 2021) and wrote an article on the impact of university mergers on the administrative careers of women and men. She recently published an article on "Feminism" in *Dictionnaire de la participation* (https://www.dicopart.fr/ 2023).

**Jorge Peixoto Freitas** has a Master's degree in Psychology and is a student on the Doctoral Programme in Human Sexuality. He is currently a researcher and project manager of the RESET: *Redesigning Equality and Scientific Excellence Together* at U.Porto. He is a certified psychotherapist with a clinical practice focusing on Systemic Intervention and Couples Therapy. His current research and clinical interests emphasise Consensual Non-Monogamy, Infidelity, and Sexual and Gender Diversity in Couple Therapy. He has worked on various research and intervention projects. An activist in equal rights and gender equality, and a member of the management board of Gentopia – NGO.

**Aleksandra Różalska**, PhD is Assistant Professor at the Department of Cultural Research, Faculty of Philology and Head of the Women Studies Centre, University of Lodz. She is a local coordinator for GEMMA: *Erasmus Mundus Master's Degree in Women's and Gender Studies*, P.I. for the Horizon 2020 RESET: *Rethinking Gender and Scientific Excellence Together* (2021–2024), and P.I. for Horizon Europe FEAST: *Food Systems That Support Transitions to Healthy and Sustainable Diets* (2022–2027). Since 2024, she has been the UL Rector's Representative for Equal Treatment. Apart from studies on media discourses and minorities in Europe and in the US, her research interests include intersectionality, media literacy, television studies, cultures of inequality, American narratives of 9/11, and media representations of refugees and migrants. She co-edited five volumes devoted to feminism in various cultural contexts.

**Agata Rudnicka**, PhD is Assistant Professor at the Faculty of Management, University of Lodz. She has 20 years of experience in teaching, training, and researching sustainability-related issues. Her main fields of interest are CSR, sustainable development, ESG, business ethics, and human rights in business. She is an author and co-author of more than 50 academic papers, books, and other materials. She has been a member of many international and national projects devoted to different dimensions of sustainability. She is actively involved in the promotion of science. She works with organisations from different sectors to guide them in sustainable transformation.

**Paulina Szymańska**, PhD, is a psychologist, therapist, and trainer. For over ten years, she has been associated with the University of Lodz, where she has worked as Deputy Director of the Institute of Psychology, as well as a researcher and a teacher. She has conducted psychoeducational lectures, training, and workshops for children, adolescents, and adults focusing mainly on different aspects of interpersonal communication, coping with stress and emotions, acceptance, tolerance, values, and teamwork. She has participated in European research and educational projects related to stress reduction (e.g., Erasmus+ VRXanny Virtual reality for anxiety and mental stress reduction in university students) or inclusive education (Erasmus+ Hey, teachers! Don't leave the kids alone!) in which she has co-designed an interactive game and training materials for educators helping to promote inclusion in different areas based on gender equality and diversification of cultural context as well as the family changes and the arrival of new types of families. She organised and actively participated in both national and international conferences. Fast learner, easy-going, and able to work with a diverse audience.

**Ruut Tikkanen,** a researcher and university teacher, INTERACT research unit, University of Oulu. Her research aims at combining education and digitality through participation, co-design, and empowerment through the lenses of

Human–Computer Interaction. She has a strong practical background on equality and diversity themes having long teaching experience in different school levels. Project management and leadership tasks have given perspective to these topics as well. In addition to computers and technology research, she is writing her PhD work in music education.

**Athena Vakali** [female] is a professor at the School of Informatics, Aristotle University, Greece, where she leads the Laboratory on Data and Web Science (Datalab https://datalab.csd.auth.gr/). She has coordinated and participated in more than 25 research projects in EU FP7, H2020, and international and national projects. She has served as a member of the EU Steering Committee for the Future Internet Assembly (2012–2014) and she has been appointed as Director of the Graduate Program in Informatics, Aristotle University (2014–2015). She has served as Program Committee co-chair for major conferences, including the ACM Gender Equality Summit (Greek Chapter) 2022, PC co-chair at the ACM/IEEE Web Intelligence Conference 2019, the EU Network of Excellence 2nd Internet Science Conference (EINS 2015), 15th Web Information Systems Engineering (WISE 2014), and 5th International Conference on Model and Data Engineering (MEDI 2015).

# ACKNOWLEDGEMENTS

The editors gratefully acknowledge the essential support of the European Commission, which funded the RESET project under the SwafS (Science with and for Society) scheme. This support was instrumental not only in enabling the work that culminated in this volume but also in advancing the values and transformative ambitions that Europe embodies – especially vital in today's often challenging and resistant academic and societal landscapes.

We extend our deep appreciation to the RESET partner teams across the seven participating research institutions, whose collaboration, insights, and commitment were fundamental throughout the project. Special thanks go to the project managers – Luce Chiodelli, Maryna Radchuk, and Marie Daviet – whose coordination and guidance played a key role in the project's success.

The active involvement of researchers and staff from the seven institutions made possible a wide range of actions and initiatives. We are particularly grateful to those who conducted fieldwork and training activities, including focus groups and surveys, and to those who contributed to the design and testing of various tools (training toolbox, GIA checklist, and GIL guidelines). Our sincere thanks also go to those directly involved in the design of institutional Gender Equality Plans, namely Gender Equality Board members as well as those who coordinated with faculties, research centres, and university services, providing essential administrative support.

Aleksandra Różalska would like to especially acknowledge the tremendous work of Dorota Golańska and Agata Rudnicka – co-authors of GEP 1.0 and 2.0 at the University of Lodz and researchers, who conducted focus groups and data analysis facilitating the process of institutional change. Your contribution to and continuous support for the RESET project made UL academia more equal and inclusive. Marion Paoletti would like to extend special thanks to Ninon Junca and Christine

Larrazet on behalf of the University of Bordeaux and Marisa Matias extends sincere thanks to Sara Isabel Magalhães and Jorge Peixoto Freitas for their significant support and valuable contributions throughout the RESET journey. Thank you for the time, expertise, and commitment you have dedicated to advancing more inclusive and excellent science.

Together, we move forward – one step closer to a truly equitable scientific community.

## AI declaration

OpenAI's ChatGPT (version 4) was used to proofread Chapter 10. This tool was used to enhance readability and ensure linguistic precision, while preserving the original content and ideas of the contributors of this book.

# SCIENTIFIC EXCELLENCE AND EQUALITY: OVERCOMING THE CONTRADICTION

*Marion Paoletti, Marisa Matias, and Aleksandra Różalska*

## Feminist critique of the notion of scientific excellence

Over the past 20 years, scientific excellence has become a watchword for academic institutions and scientific staff alike, who are now engaged in competition at local, national, European, and global levels. Researchers' and teachers' performance is measured mainly by three criteria: quantification of publications, participation in research projects, and internationalisation (Gallot et al., 2024). International rankings, first and foremost the Shanghai ranking (which dates back to 2003) exacerbate global competition between institutions to obtain more funding and attract the brightest students and researchers. In the name of excellence, they have helped to reconfigure the academic sector by concentrating funding on a few institutions.

Excellence is a vague notion, embodied in the public action for institutions and researchers, based on competitive calls for projects that "transform . . . the forms of promotion and contribute . . . to the evolution of judgement criteria and of what defines what is excellent" (Musselin, 2017, p. 27). Excellence, as a categorical imperative for staff and institutions, has been the subject of increasing debate for several years. Feminist criticism is crucial in questioning the notion of scientific excellence (O'Connor et al., 2020). The myth of meritocracy and the belief in scientific excellence are key factors contributing to inequalities in European academic institutions (Jenkins et al., 2022). In particular, feminist theories and research tend to challenge this belief: on the one hand, they criticise the processes of formal and informal power within academic organisations and of knowledge production (Benschop, 2021). On the other hand, they show that scientific excellence is a social construct projecting a disembodied academic ideal, whereas multiple intersectional inequalities are deeply rooted in its construction and evaluation.

DOI: 10.4324/9781003564348-1

Based on this feminist literature and the concrete experience of the RESET project, this book aims to answer two main questions: how can the requirement of merit, conveyed by the notion of scientific excellence, be reconciled with the existence of the multiple inequalities spread through society and reproduced by university organisations? Under which conditions are Gender Equality Plans effective when implemented in European universities in parallel with the strengthening of the excellence imperative?

Women seem more equipped to play the game of a competition based on clear criteria. More and more women are enrolled in higher education; the majority of people holding a Master's degree are women; and around half of PhD degree holders are female as well. In addition, under pressure from the European Union, equality policies and support for gender mainstreaming in science have been introduced in HEIs, whereas the notion of excellence has only been associated with the idea of "exemplary" employers. While the European Union promotes competition that is intended to be transparent, it is also, and increasingly so, the bearer of a demand for professional equality, and more and more from an intersectional perspective (see Maxime Forest's chapter in this book).

However, research on scientific excellence tends to show that the main consequence of excellence policies is to renew the famous Matilda effect highlighted by Rossiter in 1993 in response to the equally famous Matthew effect highlighted by Merton in 1968 (Rossiter, 1993). The Matilda effect describes the invisibility and undermining of women's work in the scientific world, whereas the Matthew effect refers to the mechanisms by which the most privileged (individuals and institutions) tend to increase their advantage. In the French context, Hugo Harari-Kermadec shows that funds, distributed through calls for "excellence projects", primarily benefit men and the exact sciences (Harari-Kermadec, 2019). Policies of excellence, embodied in calls for projects and rankings, accentuate gender, social, and racial inequalities. Even if the criterion of internationalisation can play a positive role for some women (Zippel, 2017), the "performance" indicators, which promote a certain model of scientific career (numerous publications, rapid and linear progression, mobility abroad, and ability to obtain funding and manage large projects), are not gender-neutral. According to Yvonne Benschop (2024), it is the very myth of excellence that is the driving force behind the reproduction of inequalities in academic organisations. For Benschop, it is nothing less than "an ideology serving the interests of the powerful", which should be rejected in the name of equality. Instead, she advocates scientific diversity rather than scientific excellence.

At the same time, the notion of scientific excellence is also a subject of criticisms relatively unconnected with those made in the name of equality. For instance, the need to protect the environment calls for a redefinition of the criteria and procedures for internationalisation. In addition, the pressure for scientific productivity, measured quantitatively, raises questions about the quality of science itself: the evaluation of science is being redefined, in particular, to take into account its impact, diversity, quality, and inclusiveness. The DORA declaration of 2012[1] was extended in September 2022 by an agreement on the reform of research evaluation at the European level.[2]

Hence, in view of the earlier, should we be arming female scientists to play the androcentric game of excellence on an equal footing with men? Should we reject the notion of excellence as an ideology serving the powerful? Should we redefine the concept by integrating social inequalities to update the notion of merit?

The answers to these questions are provided in this volume by a group of researchers representing different disciplines, who have been working together since 2020 within a European project consortium, with the aim of driving cultural change within their universities.

The RESET project's position is that we need to counteract discrimination within HEIs through specific actions that take into account inequalities persisting therein, while at the same time focusing on redefining the notion of scientific excellence to make it a more inclusive and hence comprehensive notion. This is a path that could be described as reformist and optimistic, in contrast to more radical and pessimistic positions that call for the notion of excellence to be abandoned in favour of diversity (Benschop, 2024).

## Sharing the experience of the RESET project

### *A common framework: a European project bringing together seven universities for redesigning equality and scientific excellence together*

RESET is a project that aimed at transforming the partner universities' approach to equality and diversity.[3] It was funded by the European Union's Horizon 2020 research and innovation programme, after having won a call within the Science with and for Society, under Grant Agreement no. 101006560. The RESET project sought two main objectives: to change the definition of scientific excellence at the European level and to integrate actions towards tackling intersectional inequalities in careers.[4]

With a budget of 3 million euros, involving seven European universities, coordinated by the University of Bordeaux and spread over four years (from January 2021 until December 2024), RESET was an ambitious project in terms of the quantity and scope of deliverables, production of new knowledge, and the number of actions to be implemented in each institution as part of the adoption of GEPs (first conceived in 2021 and reviewed in 2024). Because of its multiple dimensions, RESET has had implications for both teaching and research as well as for various scientific and administrative activities (particularly within Human Resources). Since its major aim was to broaden the definition of excellence and the profiles of researchers recognised as excellent, the undertaken tasks were necessarily multifaceted and cross-disciplinary. Initiatives also needed to be effective in mobilising various communities within universities, at a challenging time when polarisation on gender issues and a certain gender fatigue are becoming more and more visible.

RESET was organised around five main objectives, set out in nine Work Packages and included 46 deliverables submitted to the European Commission by the end of 2024.

The objectives of the project were the following:

1) Accomplishing change through Gender Equality Plans.
2) Training to raise awareness, empower, act upon bias, and debunk myths.
3) Changing the institutional cultural framework to anchor equality and diversity.
4) Changing the structure through the co-design of new processes.
5) Supporting the inclusion of gender in research activities.

The project was also based on two methodological principles, which were the underlying perspectives in all RESET initiatives, documents, reports, and tools.

*Co-design method*

Mobilising various actors and groups is crucial at a time of polarisation of opinions in Europe on the issues of gender, equality, and diversity, particularly among young people (Alexander et al., 2024). Therefore, it seems of the utmost importance to build common grounds and consensus within universities on equality and diversity while indifference prevails within the university (by the majority of the community) and even hostility (by a minority). Workshops, training sessions, games, and practical exercises that promote the subjectivity of each individual in academia are helpful in engaging different stakeholders and offer a favourable ground for the successful introduction of DEI culture. Co-design has been central to the processes of institutional change and knowledge production, as well as to actions undertaken both within the consortium and by each local team. It requires both preparation and leadership skills that take into account local and national contexts of the specific universities (see Netta Iivari, Eva Durall Gazulla, and Ruut Tikkanen's chapter in this book).

*Intersectionality*

The project focuses on gender relations; however, we approach the gender dimension through intersectional lenses, taking into consideration overlapping and codependent social relations and identities. Considering the increasing complexity of equality regimes in Europe and the need for recognition of multiple overlapping systems of exclusion (Kriszan et al., 2012), an intersectional perspective on inequalities within universities is necessary and timely. This is a methodological challenge that was not fully met in RESET. Using the intersectional approach in the RESET project resulted in the greater appreciation of diversity and minority groups as well as in addressing inequalities in a more inclusive way, for example, in the RESET Gender Equality Plans. While this intersectional perspective was used to design action plans adopted already in

2022 in the four RESET universities (GEPs 1.0), a number of difficulties were identified. They included the lack of available data, obstacles in appealing to all target groups, and potential foci of resistance. These have been considered while implementing intersectional measures (for example, with regard to gender and age) in the new versions of GEPs signed in 2024.

Although the seven European universities involved in the RESET project share the same understanding of cultural change needed within HEIs and similar methods to enforce it, they do not stand on the same stage of equality (whether at the level of public action or social realities). Four universities designed two successive GEPs during the course of the project: the University of Bordeaux (UBx), the University of Lodz (UL), the University of Porto (UPorto), and the Aristotle University of Thessaloniki (AUTh). Two universities have played a mentoring role due to their long-standing work on gender equality and diversity: the University of Oulu (UOulu) and the Ruhr-University Bochum (RUB), while the seventh partner, Science Po Paris (ScPo), acted as the project's internal evaluator. As the RESET project focused mainly on developing GEPs for implementing partners, it is important to consider their local contexts.

### *Local and national differences*

Beyond common goals shaped by the European project, different universities are marked by more or less favourable national frameworks concerning the imperatives of scientific excellence or the implementation of equality policies, as well as diversified local contexts. This is particularly true for the four partners implementing GEPs, whose experience is at the heart of this book – HEIs from Greece, France, Poland, and Portugal. According to the most recent EIGE report (European Institute for Gender Equality, 2024), in 2024, the countries of the four implementing partners stand far apart in terms of the Gender Equality Index. France is the only country above the EU average, ranking 6th, Portugal ranks 15th, Poland 18th, and Greece ranks 25th.

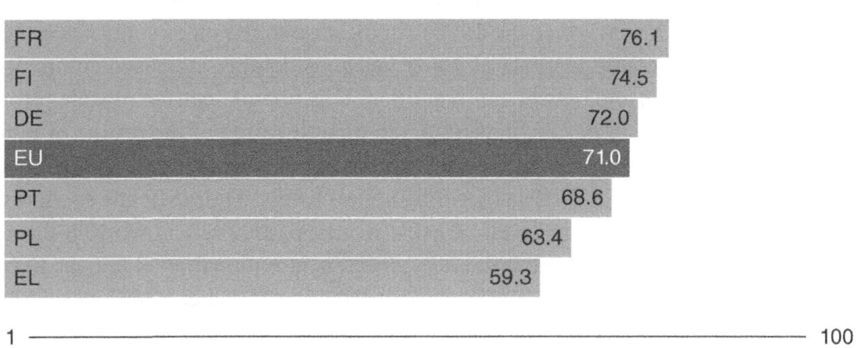


**FIGURE 0.1** Gender Equality Index (European Institute for Gender Equality Index, 2024)

Although the four countries recognise gender equality as a constitutional objective, their different rankings in the EIGE index reflect the degree of progress made in equality policies, particularly in the academic sector. For example, before the project began, the issue of career inequalities in research had already been identified as a public problem in France (with, for example, the compulsory inclusion of women in selection committees by the implementation of quotas). France also had quite a strong legislative framework in terms of sexual harassment, with incentives to set up monitoring units in each university from 2019. Since 2019, Greek universities have been legally obliged to set up Gender Equality Committees (GECs), whose role includes, among others, combating gender-based violence. This urge for action against sexual harassment by the Ministry of Science in Portugal only occurred in 2022, when U.Porto signed its first GEP.

While gender and equality issues are increasingly polarising public opinion across Europe, the strength of this political division varies from country to country. For example, at the start of the project in 2021, in Poland, the national context was marked by a strong anti-feminist backlash and resistance towards any regulations on gender equality. In the mainstream public discourse, feminist postulates have been typically framed as ideological or marginal, and the field of gender studies is not considered "real" science.

Since academic sectors in different countries are not subject to the same standards of gender equality, they are affected differently by the imperative of scientific excellence. As Part I of this book elaborates, the rise of the debate on scientific excellence is linked to European policy, but also to international university rankings and the development of New Public Management, which has affected European countries at different rates (Musselin, 2017). Established in 2003, the Shanghai Ranking of the world's 500 "best" universities was an immediate and dazzling success, particularly in France. By objectifying the hierarchy of researchers and HEIs, which was previously either informal or non-existent, it created the conditions for the university market. It is based on a measure of scientific productivity (number of publications, citations, scientific prizes, etc.) without including disciplines such as the humanities or social sciences. At the same time, the development of the New Public Management (NPM) since the 1980s (in the case of the Anglo-Saxon countries), since the 1990s (in the case of the other countries), and in Greece since the economic crisis of 2008, has considerably increased competition between researchers and between universities. This liberal doctrine of state reform, based on a reduction in public services and a quantification of the results of public bodies and public employees, has helped to configure scientific excellence at both researcher (individual) and HEIs (organisational) levels. The NMP has encouraged a "merger mania", in Kitchener and Gask's words, particularly at work in the academic sector in France, where university mergers at national level appear to be a response to the shock of the first Shanghai Ranking (Kitchener & Gask, 2003). According to a report by the European University Association, France ranked first in Europe for the number of mergers carried out between 2000 and 2015. The idea of pooling skills and resources to make universities more visible in the global competition is predominant. Moreover, the French government has allocated additional funds to nine universities awarded the Initiative of Excellence (Idex) label, through a specific programme and call for projects. The University of Bordeaux is fully in line with this national policy of excellence: as a result of a merger of universities in 2014, this university was awarded the Idex label in 2011, which was renewed in 2016.

National differences in equality policies, excellence policies, and the degree of consensus on gender and diversity issues of course partly shape the disparities between universities. All RESET universities are affected by the competition introduced by international rankings, with a tendency to concentrate funding via calls for projects and by successive reforms to evaluation systems of public research.

### *A common challenge: the gender of excellence*

The seven universities/institutions cooperating in RESET are recognised as "excellent" in their countries, either because they have a national label or because they appear at the top of national rankings. At the same time, they are plagued by inequalities in careers and positions, particularly between men and women. For a detailed presentation of local university contexts, please consult the "survey data reports" on the RESET website (the deliverable D1.2 produced under the responsibility of the RESET team at the University of Porto). This tension between aspiration to scientific excellence and inequality is particularly relevant to the four universities implementing a GEP, whose experience is at the heart of this book: the Aristotle University of Thessaloniki **(AUTh)**, the University of Bordeaux (**UBx**), the University of Lodz **(UL)**, and the University of Porto **(U.Porto).**

These four universities together employ almost 17,000 people and have almost 190,000 students.

As numerous studies have already shown, gender has a decisive impact on the types of jobs held in universities (Jenkins et al., 2022; Rosa & Clavero, 2022). Whatever the differences between countries in the academic job market, recruitment methods for scientific jobs (e.g., local or national) or status (people may exclusively be researchers, teachers, or lecturers, others may be teachers-researchers), career inequalities between women and men can be observed.

In three out of the four universities, scientific jobs are primarily held by men and in all four universities administrative jobs are held by women, with the exception of UL where more women are in academic positions (Figure 0.2).

Furthermore, the quantitative data collected from each university and available for consultation via the RESET toolkit[5] show that the gender distribution across the career ladder, particularly for scientific posts, is in line with the European trend (She Figures, 2025). The glass ceiling (or even lead sky) or "leaky pipeline" phenomena are at work in all four universities: the higher up the scientific hierarchy, the fewer women there are. At MA level, there is a majority of women, their

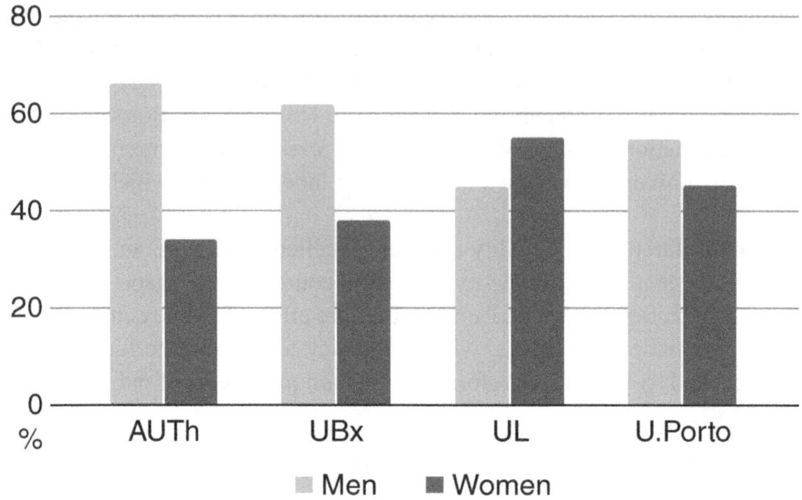


FIGURE 0.2 Sex ratio for academics per implementing partner institution, 2021

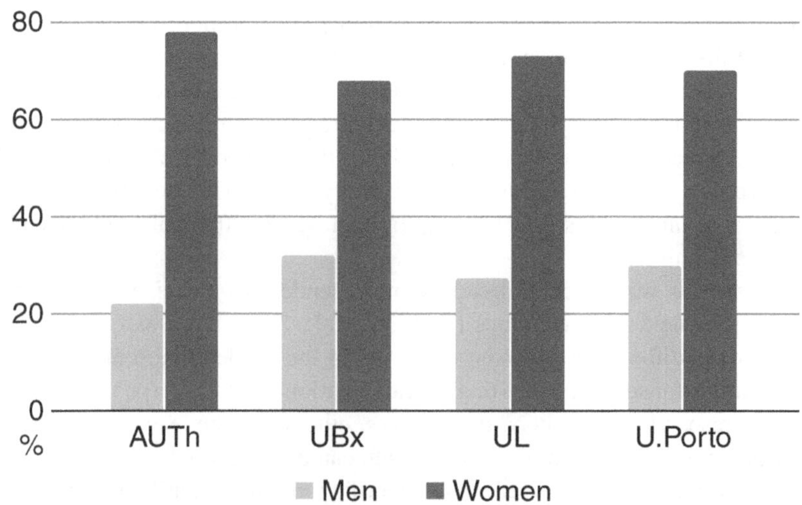


FIGURE 0.3 Sex ratio of staff per implementing partner institution, 2021

proportion decreases at doctoral level compared with Master's level, even though female thesis enrolment has increased significantly in recent years, and the proportion of women is even lower at the point of tenure and at the position of full professor. For example, in 2022, men represented around 73% of rank A in AUTh, UBx, and U.Porto, and slightly less in UL (68%). On rank D, they represented 40%

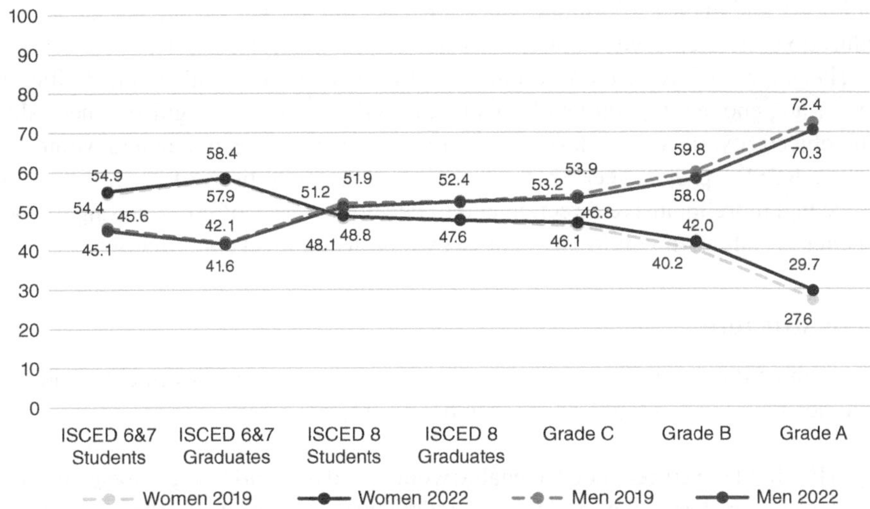


**FIGURE 0.4** Proportion (%) of men and women in a typical academic career, students and academic staff, EU, 2019 and 2022 (She Figures, 2024)

(AUTh), 56% (UBx), 53% (UP) and 36% (UL). The diagram of academic trajectories by gender produced by *She Figures* captures the phenomenon of evaporation.

The "sticky floor" phenomenon is still at work, which means that after obtaining a PhD degree, women's academic careers decelerate. There is a disproportion between the number of women with lower academic degrees (MA/MSc and PhD) – they constitute 59% of employees – and the number of women with higher degrees (associate professorship and full professorship): among the latter group only 31.5% are women. Furthermore, women are under-represented in lower level leadership positions; more women than men work in teaching-only and technical positions that are often undervalued, while men constitute a majority in more prestigious research positions.

A survey carried out in 2021 (see Part II of this book) provides some interesting insights as senior management positions are generally seen as unobtainable for both men and women, staff and academics. 32.3% of female researchers who took part in the survey felt they were not senior enough to apply for such positions; however, on a more positive note, only 6.5% of women think they lack the necessary skills for such positions.

These segregation processes are not only vertical but also horizontal, that is, women are more represented in disciplines considered to be feminine (educational sciences, psychology, etc.) and men are more represented in disciplines considered to be masculine (computer science, engineering sciences, etc.). Moreover, these areas hold an implicit hierarchy of disciplines (and funding) (Harari-Kermadec, 2019).

This European distribution of academic staff according to discipline and type of position is to be found in our universities, in a somewhat attenuated form. For example, women are a minority in engineering sciences (except at the University of Porto), and in these disciplines, they are more often teachers than researchers. In education, they represent a majority of staff, including as researchers.

The impact of gender on current scientific careers and within universities is important, and the imperative of scientific excellence, under the guise of neutrality and objectivity, runs the risk of accentuating the gaps between men and women.

The RESET project seeks both to challenge the definitions of scientific excellence in a more inclusive direction and to develop a series of cross-cutting actions to alter the situation within universities by mobilising local communities.

## Book structure

The aim of this book is to share RESET experiences, the achievements, and difficulties it has encountered. More specifically, its goal is to:

1) Highlight the current need to enable women and minorities to compete on equal terms, simultaneously moving the androcentric notion of scientific excellence in a more inclusive direction;
2) Promote the most relevant actions in the context of inclusive Gender Equality Plans with the overall objective of producing science that is a source of innovation;
3) Analyse why there is resistance to challenge the ideology of merit and promote equality and intersectionality within universities, while providing levers for action to deal with obstacles.

Written by academics who have played an operational role in promoting cultural change in their universities, this book aims to combine theory and practice, paying attention to local experiences and concrete examples. It seeks to share a number of innovative aspects:

- Discussion of the issue of scientific excellence in relation to the ambitions of European universities towards gender equality;
- Combination of a theoretical dimension informed by social sciences with the experience of concrete action plans implemented by RESET stakeholders;
- Update on the impact of European policies by taking into account the recent requirements of the Horizon Europe programme, namely the obligation to have a Gender Equality Plan, to show gender and diversity dimensions' integration into research, dissemination, and embracing of intersectionality, using a comparative perspective of several European universities;
- Analysis of the sources of support for and resistance to redefining scientific excellence and disseminating the notion of intersectionality in universities;
- Availability of resources for academics necessary for bringing about a sustainable cultural change within their organisations.

The first part, edited by Marion Paoletti, is devoted to the notion of scientific excellence. This part covers four different aspects: 1) it looks back at the notion of scientific excellence based on feminist literature in social sciences, 2) it traces the compatibility or incompatibility of European policies on scientific excellence and gender equality, 3) it takes a closer look at European academic employees by reconstructing their perception of scientific excellence, and 4) it presents the joint statement of the RESET institutions' top management in favour of a redefinition of scientific excellence accompanied by new indicators.

The second part, edited by Marisa Matias, demonstrates how universities can be mobilised in practical ways to achieve more inclusive excellence, with a focus on diversity and equality. Divided into four chapters, it first describes the co-design methodology required for mobilisation. It then examines the steps (and difficulties) of crafting action plans that extend beyond equality, aligning with Horizon Europe's expectations. The third chapter focuses on the collection and uses of equality, diversity, and inclusion (EDI) data, emphasising its crucial role in empirically and realistically supporting an intersectional approach. Finally, the fourth chapter presents a practical tool for driving institutional and cultural change by highlighting the importance of communication strategies tailored to promote equality and diversity.

The third part, edited by Aleksandra Różalska, is dedicated to the importance of training and capacity building in enhancing institutional change with regard to equality and diversity. Attention is given to designing a comprehensive and successful training programme and to the stakeholders that partake in this process. It also elaborates on RESET's take on building Communities of Practice, Gender Impact Assessment, and introducing gender and diversity dimensions in research. The four chapters in this section cover: 1) the role of trainers in GE and diversity training, 2) the role of stakeholders and their particular needs with regard to GE and diversity training, 3) the role of CoPs in enhancing a more inclusive institutional change, and 4) the transformative role of GIA tools in knowledge production.

This book emerges from the RESET project's aspiration to redesign scientific excellence through gender lenses, while advocating for a more inclusive and equitable academic landscape. By engaging with feminist and intersectional perspectives, it challenges the long-standing values embedded in traditional notions of merit and success in academia. The book is more than a theoretical exploration: it is a practical guide informed by the experiences and actions of RESET partners.

We invite readers to learn more about the opportunities and barriers in implementing cultural change within RESET institutions, and in this way, equip others to redesign excellence and promote a more diverse and inclusive academia.

## Notes

1 https://sfdora.org/read/read-the-declaration-french/.
2 https://research-and-innovation.ec.europa.eu/system/files/2022–07/rra-agreement–2022.pdf.
3 https://wereset.eu/.
4 Our sincere thanks go to Marie Daviet, project manager of the RESET project between 2023 and 2024, both for her contribution to the success of the project itself and for her help with this book.
5 https://toolkit.wereset.eu/#/.

## References

Alexander, Amy, Charron, Nicholas, & Off, Gefjon. (2024). For every action a reaction? The polarizing effects of women's rights and refugee immigration: A survey experiment in 27 EU member states. *European Journal of Political Research, 63*(4).

Benschop, Yvonne. (2021). Grand challenges, feminist answers. *Organization Theory, 2*(3). https://doi.org/10.1177/26317877211020323

Benschop, Yvonne. (2024). Un plaidoyer pour la diversité plus que l'excellence. *Travail, Genre et Société, 1*(51), 161–165. https://doi.org/10.3917/tgs.051.0161

European Commission: Directorate-General for Research and Innovation. (2024). *She figures 2024: Gender in research and innovation: Statistics and indicators*. Publications Office of the European Union. https://data.europa.eu/doi/10.2777/592260.

European Institute for Gender Equality. (2024). *Gender equality index 2024*. https://eige.europa.eu/gender-equality-index

Gallot, Fanny, Paoletti, Marion, & Pochic, Sophie. (2024). L'excellence scientifique: piège ou opportunité pour les femmes. *Travail, Genre et Société, 1*(51), 157–160. https://doi.org/10.3917/tgs.051.0157

Harari-Kermadec, Hugo. (2019). *Le classement de Shanghai. L'université marchandisée*. Le Bord de l'Eau.

Iivari, Netta, Tervo, Erkki, Käsmä, Marjuka, & Heikkinen, Mervi. (2023, June 28). Participatory design meets gender equality at European higher education institutions. https://doi.org/10.1080/15710882.2023.2215742

Jenkins, Fiona, Hoenig, Barbara, Weber, Suzanne Maria, & Wolffram, Andrea. (Eds.). (2022). *Inequalities and the paradigm of excellence in academia*. Taylor & Francis.

Kitchener, Martin, & Gask, Linda. (2003). NPM merger mania: Lessons from an early case. *Public Management Review, 5*(1), 19–44.

Kriszan, Andrea, Skjeie, Hanne, & Squires, Judith. (Eds.). (2012). *Institutionalizing intersectionality: The changing nature of European equality regimes*. Palgrave Macmillan.

Musselin, Christine. (2017). *La grande course des universités*. Presses de Sciences Po.

O'Connor, Pat, et al. (2020). Micro-political practices in higher education: A challenge to excellence as a rationalizing myth? *Critical Studies in Education, 61*(2), 195–211. https://dx.doi.org/10.1080/17508487.2017.1381629

Rodrigo Rosa, & Clavero, Sara. (2022). Gender equality in higher education and research. *Journal of Gender Studies, 31*(1), 1–7. https://doi.org/10.1080/09589236.2022.2007446

Rossiter, Margaret W. (1993). The Matthew Matilda effect in science. *Social Studies of Science, 23*(2), 325–341. https://doi.org/10.1177/030631293023002004

Zippel, Kathrin. (2017). *Women in global science: Advancing academic careers through international collaboration*. Stanford University Press.

# PART I

# Issues of compatibility

# INTRODUCTION TO PART I

## Issues of compatibility in equality and excellence

The first part takes a close look at the concept of scientific excellence, its tensions with gender equality in higher education institutions, and recent efforts to make scientific excellence and equality compatible.

This section opens with the chapter by Pat O'Connor, who summarises the existing literature on this tension. Although scientific excellence (often defined by scientific productivity) is assumed to be gender-neutral, the reality is different. Higher education institutions and scientific careers are deeply gendered, particularly in highly masculinised disciplines. The chapter sheds light on the reasons behind this "false neutrality", and expresses caution regarding new instruments designed to move beyond strictly quantitative evaluations of scientific output (e.g., the DORA or Athena SWAN declarations). Maxime Forest's chapter complements this discussion by analysing the evolution of European policies over the last three decades. On the one hand, policies have promoted scientific excellence within the European Research Area using criteria intended to be objective. On the other hand, gender equality policies incorporating feminist critiques have also gained traction. Forest's assessment of these two concomitant policies is mixed. While progress has been made, particularly with the emergence of the concept of "inclusive scientific excellence" in recent European programmes, challenges remain, especially in integrating intersectionality into these frameworks. Marion Paoletti's chapter looks at how university workers perceive the notion of scientific excellence, based on focus group data from four universities. Her findings reveal a widespread "fatigue" with regard to this imperative. While a minority of participants hold to the status quo as a guarantee of equality, the majority view the current model of excellence as primarily shaped by male gender norms. For many, ensuring equality of opportunity

DOI: 10.4324/9781003564348-3

through inclusive policies is seen as necessary for enhancing the credibility and diversity of scientific excellence. Finally, Ninon Junca's chapter describes the work undertaken by the RESET consortium in drafting a joint declaration advocating for inclusive scientific excellence. The chapter also details the fields targeted by this ambition and the innovative indicators developed to support it.

Overall, this section outlines the steps taken by feminist scholarship, European policy, and individual institutions to address the contradictions between scientific excellence and gender equality, offering pathways to make the two compatible.

# 1

# IS SCIENTIFIC EXCELLENCE A GENDERED CONSTRUCTION?

*Pat O'Connor*

## Introduction

The concept of excellence in higher education (HE) is rooted in the idea of meritocracy, meaning that "access to resources and power is granted to those who deserve it. In the context of universities, meritocracy is translated into excellence" (LERU, 2018, p. 6). Excellence can be "conceptualised as a component of the ideological apparatus of managerialism" (Santiago & Carvalho, 2012, p. 513), which in turn is the "organisational arm of neoliberalism" (Lynch, 2014, p. 968; see also Lynch et al., 2012). Thus, what purport to be objectively defined key performance indicators (KPIs) are driven by the "pursuit of marketised conceptions of excellence". They reflect a particular view of the purpose and value of higher educational institutions (HEIs) as a producer of knowledge and innovation for industry, with high-ranking research publications being the defining metric of excellence. This ignores a very different view of the nature and purpose of HEIs: one that is focused on students and teaching. However, these activities are devalued in the research excellence model and often allocated to precarious, junior, or devalued "workhorse" academic staff (such as women) (Williams & Ceci, 2015; Carlsson et al., 2021).

HEIs are perceived by many of their senior leaders as gender-neutral organisations (Van den Brink, 2015): with organisations with an explicit commitment to meritocracy being particularly susceptible to gender bias (Castilla & Benard, 2010). The gendering of HEIs has long been identified (EC, 2004; ETAN, 2000; EU, 2012). They are seen by Connell (2024) as created by men for male advancement, overwhelmingly adopting a natural science model of objectivity rooted in a decontextualised experimental methodology. There is evidence that men are more likely than women to attain the rank of full professor, and more quickly (Harris et al., 2024; Fox, 2020).

DOI: 10.4324/9781003564348-4

Most attention has focused on the gendering of the structures and culture which produce these outcomes including the formal criteria (Orupabo & Mangset, 2022), the recruitment/promotional processes and practices (O'Connor, 2020, 2023) with informal micropolitics including homosociability affecting workload allocation, the favouring of men, and the devaluing of women (Boivin et al., 2024; O'Connor et al., 2017). Gendered patterns, frequently reflecting gender stereotypes, appear to have changed little over time (Ibarra et al., 2010; Morley, 1999; Teelken et al., 2019).

Less attention has been paid to the extent to which those involved in these processes have responsibility for teaching and hence may have a real interest in broadening the concept of excellence. This may help to explain the enthusiasm emerging in Scandinavian HEIs for hiring female staff.

## Theoretical perspective

In HEIs, individual excellence is the basis for accessing positions and rewards (Scully, 1997, 2002). Thus, excellence is a legitimating discourse with the implicit suggestion that what is defined as excellent is unambiguous and objective. This has been challenged (Nielsen, 2016a, 2016b; O'Connor & O'Hagan, 2016; Van den Brink & Benschop, 2012a, 2012b). Indeed, when organisations are legally required to change and/or it is in their financial interest to do so, changes occur (O'Connor & Irvine, 2025).

Given the dominance of managerialism, little attention has been paid to the implicit devaluing of teaching and its implications for the nature and purpose of HEIs; nor to the devaluing of other activities stereotypically allocated to women, such as pastoral care, service, or lab work (Fox, 2020). Gender representation in HEIs, differential informal access to those in power, and the allocation of research and high profile "outward-facing", research activities to men and "inward-facing", "housekeeping" (Heijstra et al., 2017) activities to women perpetuate gender inequalities. These are in Valian's (2005) terms the molehills that ultimately create metaphorical mountains.

The construction of excellence has been implicitly challenged by some Norwegian studies which have included those with teaching responsibilities in the recruitment/promotion process. Thus, many of the heads of departments in Lagesen and Soboticki's (2024) study were critical of narrow definitions involving research excellence since they struggled to cover their department's teaching responsibilities if they hired staff who proved to be inadequate or unwilling teachers. In an experimental study, overwhelmingly involving academic evaluators in Norway and Sweden, the CVs of female candidates for associate professorships were seen as more competent and more hireable than their male counterparts (Carlsson et al., 2021). However, in Norway, inhouse women, even at professorial level, were hired to do the more mundane, low profile, less prestigious academic housekeeping activities that were essential to keep departments running

(Moratti, 2021, p. 927), with the internal bonus for women candidates reflecting and reinforcing their lower value.

Attention will focus on the content of excellence, as well as the underlying gendered processes. First, however, we look at the characteristics of those seen as excellent.

## The characteristics of those who are seen as excellent

Excellence has been viewed as reflected in the desirable characteristics of those in powerful positions, whether as Rectors/Presidents, Vice Presidents, or full professors. With a small number of notable exceptions, men make up the majority of those in these positions: "doing man" being conflated with "doing power" (Foley & Williamson, 2019). Gender stereotypes are particularly evident at senior management and professorial level where "think manager, think male" persists (Schein et al., 1996; Van den Brink & Benschop, 2012b). With managerialism, those who are in such senior management positions also tend to be excellent researchers and play a key role in many HEIs in perpetuating research excellence as the most important criterion in recruitment/promotion, especially to the most prestigious positions.

There has been a widespread recognition of the "greedy" nature of HEIs with both men and women academics working very long hours (Misra et al., 2012; Winslow, 2010). The gendered monastic stereotype underpinning science, technology, engineering, and maths (STEM) has also been recognised (O'Connor, 2014; White, 2014) and is reflected in negative attitudes to maternity leave, women's embodiment, and non-work activities (Fitzgerald, 2018). Access to permanent positions in HEIs (e.g., through the tenure track process) frequently coincides with the most intensive childbearing and rearing period (O'Connor & Drew, 2023). Academic women are less likely than their male counterparts to have children (Misra et al., 2012). Winslow (2010, p. 788) found that parenthood was a disadvantage for women but that it advantaged men, reflecting as it did "the male ideal worker norm". Many German HEIs are more comfortable attributing gender under-representation to a lack of "family friendliness", rather than to the gendering of science, criteria, procedures, and processes or the devaluing of women/female disciplines (Zippel et al., 2016). Among Danish social scientists, there was a 24% disadvantage in advancement for women within the first seven years after their PhD, with publishing, parenthood, and parental leave accounting for approximately 40% of that gender gap. This highlighted the continued importance of looking at "social dynamics *within* organisations" (Nielsen et al., 2024, p. 19).

On looking at the perpetuation of sameness in university leadership in elite universities at the Rector/Presidential level, implicitly intersectional approaches have occasionally been adopted. Thus, Aiston and Fitzgerald (2023) highlight incumbents' routes from undergraduate level in highly ranked, high-prestige universities to rector/presidential positions in them. Like Carvalho and De Lourdes Machado-Taylor (2017), they suggest that some variation in gender can be tolerated so long as those who reach these positions are a "cultural fit" and hence likely to perpetuate the status quo.

Geographical mobility is required by the EU, with men more likely than women to be mobile (Harris et al., 2024; EC, 2025) – not least because of their lesser caring commitments and the societal expectations that a female spouse will prioritise her husband's career. The Spanish HEI system characterised by loyalty and inbreeding (Sanz-Menéndez et al., 2013) is an exception, but still perpetuates male dominance (Montes Lopez & O'Connor, 2019). In computer science in the US, the strongest predictor of good/excellent chances of promotion was collaboration in the previous three years on publications/proposals with those in their home unit (Fox, 2020), thus implicitly suggesting limits to the importance of geographical mobility.

In summary, the characteristics of those who are perceived as excellent include maleness and the presence of children (for men, but not women), with geographical mobility being in an ambiguous position.

## Content of the evaluative criteria

Excellence is often simply used as a "hooray word" (Whyte, 2005) in policies and effectively tautologically defined. In the UK in the Research Evaluation Framework, references to "excellent research/researchers" were used as an "empty signifier" (O'Connor & Barnard, 2021, p. 55). Thus, evaluation panels were encouraged to define it in particular disciplinary contexts, and the possibility that such definitions might involve gendered biases was ignored. It was also devoid of content in internal promotion documents in a UK HEI, beyond referring to "originality, significance, and rigour", with no indication as to what the appropriate indicators of these might be nor how they might vary at different levels of the academic hierarchy. This opens the way for gender bias, and in the context of masculinised HEIs, such bias is very unlikely to favour women (LERU, 2018; Fox, 2020). The "essentially contested" nature of the concept of excellence and the indicators of it was recognised by those who had created the Science and Technology Indicator used to rank the performance of the member states and to distribute funds (Ferretti et al., 2018, p. 732; Campbell, 2018).

Theoretically in HEIs, there are three dimensions to excellence, reflecting the mandate of higher education, viz., teaching, research, and service (whether to the organisation, the profession, or wider society). However, in many Western universities, research has become the most important criterion for recruitment and advancement, particularly in the most prestigious HEIs, and/or where the majority of those on the recruitment/promotional panels are senior academics with a research focus. Unusually, in Norwegian HEIs, the process includes not only a scientific committee of experts (who are likely to be senior academics) but also an internal interview committee including student representatives, with the final recommendation being made by the HOD. Many of the HODs (both men/women and from STEM/Humanities) were doubtful about the objectivity of what purported to be meritocratic procedures. They were sensitive to departmental "fit", and to their departments' teaching responsibilities in a context where, if a male and female candidate were perceived as equal, they could recommend the appointment of a woman (on the grounds of "moderate gender quotas": Lagesen & Suboticki, 2024, p. 694).

In a cross-national STEM study, departmental "fit" and broader definitions of "excellence" were also referred to by respondents in Turkish, Irish, German, Italian, and Bulgarian HEIs, with those in Germany and Turkey adverting to a tension between departmental priorities and senior management's focus on research excellence. Difficulties in defining excellence objectively were recognised; with the differentiation between adequacy and excellence highlighted. Thus, a respondent from Bulgaria noted, "I have never come across an exact definition, a scientific one, of this term. Maybe the Nobel Prize . . . but . . . how many Nobel Prize winners are there?" (O'Connor et al., 2017, p. 205). Of course, "local fit" could facilitate the maintenance of a predominantly male gender profile, with relational practices such as sponsorship and "paying forward" being part of the assessment (O'Connor et al., 2017).

Within most Western HEIs, although various indicators of research productivity may be prioritised, they all, to varying degrees, can be quantified, and hence are attractive as measures of "excellence" in a natural science paradigm. It can be seen as reflecting a managerialist concern with global ranking (Hazelkorn, 2015). This focus serves to implicitly advantage men, since women are more likely than men to spend a higher proportion of their weekly work time teaching (Winslow, 2010). That can be partly explained by their disproportionate employment in teaching-intensive institutions where there are less resources for research (Fox, 2020) and their over-representation at the lower levels of the academic hierarchy (Santos & Dang Van Phu, 2019). Winslow (2010, p. 788) found that women were more likely than men to be teaching more than they preferred, with men more successful in getting their preferences met over and above their other measurable (gendered) advantages including rank and type of institution. The impact of COVID-19 on women's research output has begun to be explored, highlighting the impossible situation of women with young children, where management expectations were that teaching should carry on, despite the disappearance of most caring structures (Nash & Churchill, 2020).

Similar difficulties around the definition of excellence arise in the evaluation of research proposals. Thus, although the European Research Council Grants (ERC) attempted to define excellence in research projects (as addressing important challenges, being ambitious, and high risk/high gain; proposing and conducting groundbreaking research; and demonstrating creative independent thinking and achievements beyond the state of the art), it was difficult to identify gender-neutral indicators, for example, being the lone or first author could be seen as an indicator of creativity but it could also reflect gendered power dynamics in the research team (Schiffbaenker et al., 2022).

Gender differences in published research output are the most common indicator of excellence and are used to explain women's under-representation in senior positions. Women do tend to have lower average publication rates than men, although these differences vary by career stage, with women more likely to be in lower ranked positions (Misra et al., 2012) and in teaching-heavy departments (Nielsen et al., 2024). However, research productivity did not explain much of the gender gap in tenure in the US (Weisshaar, 2017). The gender gap also varied between disciplines (being largest in English and smallest in Computer Science). Even after accounting for individual research productivity, and the ranking of their original department, women, on average, were likely to get tenure in lower ranked departments, resulting in the conclusion that the "'leading culprit' for perpetuating that gender gap in getting tenure was the evaluation process itself" (Weisshaar, 2017, p. 554). Sheltzer and Smith (2014) found that male faculty members tended to employ fewer female graduate students and postdocs than female faculty members did, and this trend was heightened among elite male (but not female) faculty.

In New Zealand, men were twice as likely as their female counterparts to be professors or associate professors; controlling for research score, age, field, and type of university, and including/excluding research productivity did not affect these results (Brower & James, 2020). Men were 24% more likely than women to be promoted to full professor in Italy although both had similar research productivity (Marini & Meschitti, 2018). Similarly, the odds of getting tenure were twice as high for men as for their female counterparts in Spanish universities (Sanz-Menéndez et al., 2013). Thus, although women tend to have lower research productivity than men, this only partially explains the gender gap.

Women also obtain lower levels of research funding than men (Husu, 2014; Schmaling & Gallo, 2023) and this affects the number of PhD students they supervise, since these are frequently funded from such grants. Authorship is also gendered, with women less likely than men (regardless of their position) to access it, particularly in highly cited articles (Ross et al., 2022). Articles that have women as first authors are significantly less likely to be cited than those in which men are first (particularly in medical science: Zhou et al., 2024; Chatterjee & Werner, 2021; Sebo & Clair, 2023). In psychology and veterinary disciplines (which are predominantly female ones in the US), the citation gender gap is larger than average (see Wu, 2024). Various explanations have been offered for this including the undervaluing of women's research; homophily; gender bias in self-citing; the size of women's networks; the nature of their research topics; and women's status/visibility as reflected in professional invitations and on social media (Chatterjee & Werner, 2021; Wu, 2024). Many of these reflect gendered patterns.

Although an attempt (for example, in the UK) has been made to develop a Teaching Evaluation Framework, it is less highly valued than the equivalent Research Evaluation Framework. Attention has tended to focus on student satisfaction and gender bias in student evaluations has been identified (Chávez & Mitchell, 2020; Fan et al., 2019). The heart of the teaching relationship is essentially a kind of caring relationship (Lynch, 2022) and hence difficult to measure. Various staff characteristics (including employment status) may also reflect gender. Thus, even if teaching is included as a criterion in constructions of excellence, the methodology used and the existence of gender bias mean that it is unlikely to favour women.

The Declaration on Research Assessment (DORA) highlighted the importance of moving beyond metrics, such as journal impact ones, and had been signed by organisations in 164 countries by February 2024. However, there is still no agreement on what should be included in DORA (Fritch et al., 2021); nor is there any evidence of its efficacy in reducing gender bias. Indeed its inherent focus on narrative may exacerbate gender inequality since gender bias exists in other narrative accounts such as letters of recommendation (Madera et al., 2019), with the proportion of such letters devoted to intellectual brilliance compared to temperament being much less for women than for men (see LERU, 2018, p. 13). Thus, DORA can be seen as another manifestation of "system-supporting popular inaction" (Tauber & Moughalian, 2022). Similarly, Athena SWAN, a popular equality charter and evaluation scheme, has had little impact on the gender profile of senior positions; gender pay gap; or micropolitical processes, such as encouraging women to apply for promotion, with junior women doing the work to achieve the Athena SWAN award and "normal" gendered processes and practices persisting (O'Connor, 2020).

## Evaluation practices

Even if the criteria are clear, the ways in which they are applied and the processes operating in the interview/evaluation panel may be gendered. Schiffbaenker et al. (2022) concluded that in the ERC evaluation of research proposals, gender bias persisted even after controlling for actual performance (see also Benschop, 2021). This may reflect the dynamics of the evaluation board, the different interpretation of criteria, or their application (e.g., the bar being set higher for women); more frequent "doubt raisers" about their performance; the shifting of criteria to support a favoured male applicant; and differentially evaluating the likeability and competence of male and female applicants (Correll, 2017).

Even where the criteria are indicated in the advertisement and used as a framework in the interview situation, practices can develop that allow gender bias to affect outcome. Thus, for example, a practice whereby a board meets immediately prior to the interviews (but after having received the applications) to review the weightings in the framework offers the possibility of "adjusting" the weightings attached to identified criteria, allowing favoured candidates to be chosen in what looks like an objective numerical system. Even within research output, in a situation where one candidate has 18 publications and another has 30, defining the cut-off point for a high score as 15 publications or more has the effect of eliminating the difference between the two candidates on this dimension (O'Connor & O'Hagan, 2016). Although decisions about candidates at the top and bottom are often clear-cut, differences between those in the middle are frequently small and can culminate in "rather arbitrary" decisions which reflect the composition of the board and the characteristics of the evaluators (Van Arensbergen et al., 2014), with gendered power dynamics including "horse trading" affecting outcome (O'Connor & O'Hagan, 2016).

The bar may also be set higher for women than men (Wenneras & Wold, 1997; Witteman et al., 2019) with identical experimental CVs being differentially evaluated, and both male and female academics favouring the male candidates' CV and at a higher salary (Moss-Racusin et al., 2012). Similarly, if a female candidate came from a prestigious research setting, the question as to whether her research reflected her own abilities was likely to be raised, whereas if a male candidate came from a similar background, this was viewed as evidence of his superior ability. Where independence was a criterion, female applicants who were not able to prove this were penalised, whereas male applicants were treated more indulgently. Similarly, a lack of mobility worked against women but not men (Schiffbaenker et al., 2022). Gendered expectations have been identified: women being required to be "extraordinary" ("sheep with five legs": Van den Brink & Benschop, 2012a), while men's deficiencies were seen as opportunities for improvement.

Gendered stereotypes may also have an impact and be reflected in a perceived "lack of fit" by women in predominantly male areas (such as science in HEIs), with competence and likeability being in tension for women but not men (Fitzgerald, 2018). In Schiffbaenker et al.'s (2022) study, the expectation of applicants was that they would "oversell" (i.e., show high levels of confidence and assertiveness). But whereas the positive evaluation of this was unequivocal in the case of men, there was a very narrow line between acceptable/unacceptable levels for women. Self-promotion is encouraged by DORA and is likely to be differentially enacted and valued in men and women (Fox et al., 2017; Vinkenburg et al., 2022).

Despite requirements that those with close ties or even a colleague from the same institution withdraw from the evaluation, nepotism also continued to affect the outcome (Wenneras & Wold, 1997), and to work more strongly for men than women (Schiffbaenker et al., 2022). It has been re-labelled as sponsorship (O'Connor et al., 2019). This involves senior managers with influence leveraging their own power and influence to advance the career of their protégés, whether by advocating, recommending, protecting, or fighting for him/her. In male-dominated structures, men are most likely to advocate for other men (O'Connor & Montes-Lopez, 2022). Sponsorship may be provided at a price, with those obtaining it "paying forward" by undertaking unpaid work in anticipation of future support. This has tended to be ignored and/or confused with mentorship, which reflects a deficit model of women (O'Connor et al., 2019).

The EC (2012) stressed the importance of gender balance (40%) on interview boards and assessment panels, assuming that it will promote gender equality. However, although this provides role models for women and increases women's understanding of how power works, it may not increase the success rate of women applicants. Schiffbaenker et al. (2022) showed that (controlling for the overall success rate on the panel) the share of female panel members had a negative effect. Similarly, bias against women applicants was even stronger among female than male evaluators, with junior female evaluators being particularly likely to favour opposite-sex candidates (Moss-Racusin, 2012, p. 16476). Thus, women in male-dominated masculinist power structures may reflect internalised misogyny. More research is needed on the conditions under which this occurs.

The subtlety of the biases of both men and women was implicitly suggested by Solga et al.'s (2023) experimental German study. There, both men and women were likely to invite women with lower (but not those with the highest) qualifications, thus conforming to gender equality expectations, while at the same time ensuring that women would not actually be appointed. Orupabo and Mangset (2022) also noted that in Norway, women (and other devalued categories) were on the initial shortlist – but were excluded at the evaluative stage when male-dominated purportedly gender-neutral criteria prevailed.

Specific disciplinary challenges were identified by Van den Brink and Benschop (2012b). Thus, in humanities, where female candidates were most plentiful, exclusionary practices were informal and reflected differential access to inner track information and sponsorship: processes that would not exist in experimental studies (Carlsson et al., 2021; Williams & Ceci, 2015). In the natural sciences where there were fewer women candidates, extra efforts were made to include women on shortlists, but women fell out at the final selection because they were not seen to conform to the stereotypical image of the ideal scientist. In medical science, although there was a substantial proportion of qualified women, very few were appointed, partly because of closed recruitment competitions (where women were not even considered) and partly because of a paternalistic stereotypical view of women as unsuitable ("too nice", "too young", etc.) for jobs that were seen to involve an element of management. A focus on the stages and disciplines where specific gender inequality practices come into play was also referred to by Solga et al. (2023) and could usefully be further researched.

Voluminous applications/CVs implicitly encourage busy academics to rely on subjective or intuitive feelings and "judgement" to differentiate between candidates, potentially relying on gendered stereotypes (O'Connor & O'Hagan, 2016; Spence, 2019; Lamont, 2009). Involvement in such evaluative activity is typically not valued and frequently now the "payback" is unclear. Hence, it is easy to see how academics might simply opt for continuing to use flawed metrics rather than spending very considerable amounts of time going through very extensive CVs or reading the applicants' work, as recommended by DORA.

## Summary and conclusions

HEIs are highly political-gendered spaces. Yet, scientific excellence, operationalised in terms of research publications, is presented as gender-neutral. It is a taken-for-granted legitimator (O'Connor & White, 2021) of the gender profile of senior positions. It has increasingly been challenged. However, there seem to be few alternatives other than those focusing on societal impact and relevance and/or on the "health" of research groups, assessed in terms of the absence of gender-based violence and harassment.

Many recommendations have been put forward regarding ways to tackle gender bias in evaluative activities including identifying a broader range of contributions (especially including teaching and service); prioritising recent achievements; using metrics cautiously; favouring quality over quantity of publications; using partially anonymised CVs at the first round and at qualitative evaluation; focusing on "academic" rather than biological age; identifying clear criteria and procedures; having external observers who can intervene on boards; ensuring that the cascade model is used (a soft quota); making the CVs of successful and unsuccessful candidates publicly available; and having clear accountability and strong leadership promoting targets or quotas (LERU, 2018; HEA, 2016; Vinkenburg et al., 2022). Although specific strategies are useful, they will not be successful without challenging the definition of excellence, including those responsible for teaching in the evaluation processes, and recognising that ultimately de-gendering scientific excellence requires challenging gendered power in organisations (O'Connor, 2023). This means revealing the implicit advantages enjoyed by many men. It means challenging the primacy attached to constructions of excellence which prioritise research publications at the expense of teaching. It also involves leveraging men's commitment to working to end gender inequality. It necessitates a recognition that current evaluative criteria, procedures, and processes are not gender-neutral. The underlying question which is rarely posed is whether, since evaluation and gender are both social processes, it is possible to create criteria, procedures, and processes which are not biased.

Insufficient attention has also been paid to disciplinary variation in the construction and operationalisation of excellence and to variation in the exclusionary strategies used at different stages in the evaluative process in particular disciplines (Van den Brink & Benschop, 2012b). However, although the practices maintaining the status quo may differ, the underlying problem remains the same: namely, the devaluation of women and the valorisation of men, while asserting that such evaluations are gender-neutral. Ultimately, those in leadership positions within such organisations who care about the future of HEIs and the creation of knowledge cannot afford to collude with gender bias with its consequent scientific and societal implications (Boivin et al., 2024). International organisations such as the EU, the OECD, and the UN have already made clear their support for a gender equality agenda. Constructing and operationalising non-gendered constructions of scientific excellence is a fundamental challenge, which demands integrity and courage as well as creativity.

## References


Aiston, Sarah, & Fitzgerald, Tanya. (2023). Trajectories of merit: Re-viewing leadership in elite universities. *Educational Review, 76*(1), 29–45. https://doi.org/10.1080/00131911.2023.2024576

Benschop, Yvonne. (2021). Sheep with five legs: Gender and excellence. *EWORA webinar: Excellence and Gender*. https://www.ewora.org/webinar-october-18–2021

Boivin, Nicole, Tauber, Susanne, Beisiegel, Ulrike, Keller, Ursla, & Hering, Janet G. (2024). Sexism in academia is bad for science and a waste of public funding. *Nature Reviews Materials, 9*, 1–3. https://doi.org/10.1038/s41578-023-00624-

Brower, Ann, & James, Alex. (2020). Research performance and age explain less than half of the gender pay gap in New Zealand universities. *PLoS One, 15*(1), e0226392. https://doi.org/10.1371/journal.pone.0226392

Campbell, P. (2018). Science needs to define excellence. *Nature, 554*(7693), 403–404. https://doi.org/10.1038/d41586-018-02183-y

Carlsson, Magnus, Finseraas, Henning, Midthoen, Arnfinn H., & Rafnsdottir, Guðbjörg Linda. (2021). Gender bias in academic recruitment? Evidence from a survey experiment in the Nordic region. *European Sociological Review, 37*(3), 399–410. https://doi.org/10.1093/esr/jcaa046

Carvalho, Teresa, & De Lourdes Machado-Taylor, Maria. (2017). The exceptionalism of women rectors: A case study from Portugal. In K. White & P. O'Connor (Eds.), *Gendered success in higher education*, 111–132. Palgrave Macmillan.

Castilla, Emilio J., & Benard, Stephan. (2010). The paradox of meritocracy in organizations. *Administrative Science Quarterly, 55*(4), 543–676. https://doi.org/10.2189/asqu.2010.55.4.543

Chatterjee, Paula, & Werner, Rachel M. (2021). Gender disparity in high impact journals. *JAMA Network Open, 4*(8), e2127739. https://doi.org/10.1001/jamanetworkopen.2021.27739

Chávez, Kerry, & Mitchell, Kristina M. W. (2020). Exploring bias in student evaluations: Gender, race, and ethnicity. *PS: Political Science & Politics, 53*(2), 270–274. https://doi.org/10.1017/S1049096520000397

Connell, Raewyn. (2024, June 7). Gender and higher education. In *Re-imagining higher education otherwise: Conversations on gender and care for more equitable futures [online]*. TUS and Sociological Association of Ireland.

Correll, Shelley J. (2017). Reducing gender biases in modern workplaces: A small wins approach to organizational change. *Gender and Society, 31*(6), 725–750. https://doi.org/10.1177/0891243217738936

EC (European Commission). (2004). *Gender and excellence in the making*. https://op.europa.eu/fr/publication-detail/-/publication/ce10dfeb-4e36–49ed-a9ec-b2182c8b986f

EC (European Commission). (2025). *She figures 2024 – Gender in research and innovation – Statistics and indicators*. https://data.europa.eu/doi/10.2777/592260

ETAN. (2000). *Science policies in the European Union: Promoting excellence through mainstreaming gender equality*. Luxembourg: Publications Office for European Communities. https://www.amites.org/sites/default/files/pdf/publicaciones/informe_ETAN_2000.pdf

EC (European Commission). (2012). *Structural change in research institutions: Enhancing excellence, gender equality and efficiency in research and innovation*. https://eige.europa.eu/sites/default/files/structural-changes-final-report_en_0.pdf

Fan, Yanon, Shepherd, Laura J., Slavich, Eve, Waters, David, Stone, Mars, Abel, R., & Johnston, E. (2019). Gender and cultural bias in student evaluations: Why representation matters. *PLoS One, 14*(2), e0209749. https://doi.org/10.1371/journal.pone.0209749

Ferretti, Federico, Guimaraes Pereira, Angela, Vertesy, Daniel, & Hardeman, Sjoerd. (2018). Research excellence indicators: Time to reimagine the 'making of'. *Science and Public Policy, 45*(5), 731–741. https://doi.org/10.1093/scipol/scy027

Fitzgerald, Tanya. (2018). Looking good and being good: Women leaders in Australian universities. *Education Sciences, 8*(2), 54. https://doi.org/10.3390/educsci8020054

Foley, Meraiah, & Williamson, Sue. (2019). Managerial perspectives on implicit bias, affirmative action, and merit. *Public Administration Review, 79*(1), 35–45. https://doi.org/10.1111/puar.13001

Fox, Mary Franck. (2020). Gender, science and academic rank: Key issues and approaches. *Quantitative Science Studies, 1*(3), 1001–1006. https://doi.org/10.1162/qss_a_00026

Fox, Mary Franck, Whittington, Kjersten Bunker, & Linkova, Marcela. (2017). Gender (in) equity and the scientific workforce. In U. Felt, R. Fouche, C. A. Miller, & L. Smith-Doerr (Eds.), *Handbook of social science and technology studies*, 701–731. MIT Press.

Fritch, Rochelle, Hatch, Anna, Hazlett, Haley, & Vinkenburg, Claartje. (2021). Using narrative CVs: Process optimization and bias mitigation. *SFDORA*. https://sfdora.org/resource/using-narrative-cvs-process-optimization-and-bias-mitigation/

Harris, Richard, Mate-Sanchez-Val, Mariluz, & Marin, Manuel Ruiz. (2024). Gender disparities in promotions and exiting in UK Russell Group universities. *Applied Economics*. https://doi.org/10.1080/00036846.2024.2361384

Hazelkorn, Ellen. (Ed.). (2015). Introduction: The geopolitics of rankings. In *Rankings and the reshaping of higher education: The battle for world-class excellence* (2nd ed., 1–20). Palgrave Macmillan.

HEA (Higher Educational Authority). (2016). *National review of gender equality in Irish Higher Education Institutions*. https//hea.ie/assets/uploads/2017/06/HEA-National-Review-of Gender-Equality-in-Irish-Higher-Education-Institutions.pdf

Heijstra, Thamar Melanie, Steinthorsdottir, Finnsborg Salome, & Einarsdottir, Thorgedur. (2017). Academic career making and the double-edged role of academic housework. *Gender and Education, 29*(6), 764–780. https://doi.org/10.1080/09540253.2017.1343324

Husu, Liisa. (2014). Research funding gap: Her excellence dwarfed by his excellence. *EuroScientist*. http://euroscientist.com/2014/06/research-funding-gap-excellence-dwarfed-excellence/

Ibarra, Herminia, Carter, Nancy, & Silva, Christina. (2010). Why men still get more promotions than women. *Harvard Business Review, 88*(9), 80–126.

Lagesen, Vivian Anette, & Suboticki, Ivana. (2024). Performing excellence and gender balance in higher education. *Higher Education, 88*, 683–701. https://doi.org/10.1007/s10734-024-00908-y

Lamont, Michele. (2009). *How professors think: Inside the curious world of academic judgement*. Harvard University Press.

LERU. (2018). *Implicit bias in academia: A challenge to the meritocratic principle and to women's careers – and what to do about it (paper no. 23)*. https://www.leru.org/files/Publications/Implicit-bias-in-academia-Full-Paper.pdf

Lynch, Kathleen. (2014). New managerialism: The impact on education. *Concept, 5*(3), 1–11.

Lynch, Kathleen. (2022). *Care and capitalism: Why affective equality matters for social justice*. Polity Press.

Lynch, Kathleen, Gummell, Bernie, & Devine, Dympna. (2012). *New managerialism in education: Commercialisation, carelessness and gender*. Palgrave Macmillan.

Madera, Juan M., Hebl, Michelle R., Dial, Heather, Martin, Randi, & Valian, Virginia. (2019). Raising doubt in letters of recommendation for academia: Gender differences and their impact. *Journal of Business and Psychology, 34*(3), 287–303. https://doi.org/10.1007/s10869-018-9591-3

Marini, Gulio, & Meschitti, Viviana. (2018). The trench warfare of gender discrimination: Evidence from academic promotions to full professor in Italy. *Scientometrics, 115*(2), 989–1006. https://doi.org/10.1007/s11192-018-2820-0

Misra, Joya, Lundquist, Jennifer Hickes, & Templer, Abby. (2012). Gender, work time and care responsibilities. *Sociological Forum, 27*(2), 300–323. https://doi.org/10.1111/j.1573–7861.2012.01343.x

Montes Lopez, Estrella, & O'Connor, Pat. (2019). Micropolitics and meritocracy: Improbable bedfellows? *EMAL, 47*(5), 678–693. https://doi.org/10.1080/02607476.2019.1637254

Moratti, Sofia. (2021). A woman's place is in the 'home'? Gender-specific hiring patterns in academia in gender-equal Norway. *Journal of Sociology, 57*(4), 916–934. https://doi.org/10.1177/1440783321995737

Morley, Louise. (1999). *Organising feminisms: The micropolitics of the academy*. St. Martin's Press.

Moss-Racusin, Corinne A., Dovidio, John F., Brescoll, Victoria L., Graham, Mark J., & Handelsman, Jo. (2012). Science faculty's subtle gender biases favor male students. *Proceedings of the National Academy of Sciences, 109*(41), 16474–16479. https://doi.org/10.1073/pnas.1211286109

Nash, Meredith, & Churchill, Brendan. (2020). Caring during COVID-19: A gendered analysis of Australian university responses to managing remote working and caring responsibilities. *Gender, Work and Organization, 27*(5), 838–846. https://doi.org/10.1111/gwao.12532

Nielsen, Mathias Wullum. (2016a). Limits to meritocracy? Gender in academic recruitment and promotion processes. *Science and Public Policy, 43*(3), 386–399. https://doi.org/10.1093/scipol/scv056

Nielsen, Mathias Wullum. (2016b). Gender inequality and research performance: Moving beyond individual-meritocratic explanations of academic advancement. *Studies in Higher Education, 41*(11), 2044–2060. https://doi.org/10.1080/03075079.2014.964240

Nielsen, Matthias Wullum, Pedersen, Jens V., & Larregue, Julian. (2024). Getting ahead in the social sciences: How parenthood and publishing contribute to gender gaps in academic career achievement. *British Journal of Sociology*, 1–25. https://doi.org/10.1111/1468–4446.13088

O'Connor, Pat. (2014). *Management and gender in higher education*. Manchester University Press.

O'Connor, Pat. (2020). Why is it so difficult to reduce gender inequality in male-dominated higher education organisations? A feminist institutional perspective. *Interdisciplinary Science Reviews, 45*(2), 207–228. https://doi.org/10.1080/03080188.2020.1760785

O'Connor, Pat. (2023). Is gendered power irrelevant in higher educational institutions? Understanding the persistence of gender inequality. *Interdisciplinary Science Reviews, 48*(4), 669–686. https://doi.org/10.1080/03080188.2023.2256591

O'Connor, Pat, & Barnard, Sarah. (2021). Problematising excellence as a legitimating discourse. In Pat O'Connor & Kate White (Eds.), *Gender, power, and higher education in a globalised world*, 47–69. Palgrave Macmillan. https://doi.org/10.1007/978-3-030-58706-3_3

O'Connor, Pat, & Drew, Eileen. (2023). The tenure track model: Its acceptance and perceived gendered character. *Trends in Higher Education, 2*(1), 62–76. https://doi.org/10.31234/osf.io/wfh9z

O'Connor, Pat, & Irvine, Gemma. (2025). Understanding Dramatic Changes in the Gender Profile of Presidents of Irish Public Universities. In L. McIlrath, H. Maher, L. Leeson, & M. Connolly (Eds.), *Transforming gender equality and inclusion within higher education in Ireland*, Chapter 10. Palgrave Macmillan.

O'Connor, Pat, & Montes-Lopez, Estrella. (2022). Excellence? In F. Jenkins et al. (Eds.), *Inequalities and the paradigm of excellence in academia*, 150–165. Routledge. https://doi.org/10.4324/9780367336761

O'Connor, Pat, Montes-Lopez, Estrella, O'Hagan, Clare, Wolffram, Andrea, Aye, Manuela, Chizzola, Valentina, Mich, Ornella, Apostolov, Georgia, Topuzova, Irina, Saglamer, Gulsun, Tan Mine G., & Çağlayan, Hulya. (2017). Micropolitical practices in higher education: A challenge to excellence as a rationalising myth? *Critical Studies in Education, 61*(2), 195–211. https://doi.org/10.1080/17508487.2020.1732561

O'Connor, Pat, & O'Hagan, Clare. (2016). Excellence in university academic staff evaluation: A problematic reality? *Studies in Higher Education, 41*(11), 1943–1957. https://doi.org/10.1080/03075079.2016.1161380

O'Connor, Pat, O'Hagan, Clare, Myers, Eva Sophia, Baisner, Liv, Apostolov, Georgia, Topuzova, Irina, Saglamer, Gulsun, Tan Mine G., & Caglayan, Hulya. (2019). Mentoring and sponsorship in higher education institutions: Men's invisible advantage in STEM? *Higher Education Research and Development, 39*(4), 1–14. https://doi.org/10.1080/07294360.2019.1623075

O'Connor, Pat, & White, Kate. (2021). Power, legitimating discourses and institutional resistance to gender equality in higher education. In Pat O'Connor & Kate White (Eds.), *Gender, power, and higher education in a globalised world*, 187–207. Palgrave Macmillan. https://doi.org/10.1007/978-3-030-58706-3_11

Orupabo, Julia, & Mangset, Marte. (2022). Promoting diversity but striving for excellence: Opening the 'black box' of academic hiring. *Sociology, 56*(2), 316–332. https://doi.org/10.1177/00380385211013143

Ross, Matthew B., Glennon, Britta M., Murciano-Goroff, Raviv, Berkes, Enrico G., Weinberg, Bruce A., & Lane, Julia I. (2022). Women are credited less in science than men. *Nature, 608*, 135–145. https://doi.org/10.1038/s41586-022-04966-w

Santiago, Rui, & Carvalho, Teresa. (2012). Managerialism rhetorics in Portuguese higher education. *Minerva, 50*, 511–532. https://doi.org/10.1007/s11024-012-9217-9

Santos, Georgina, & Dang Van Phu, Stephanie. (2019). Gender and academic rank in the UK. *Sustainability, 11*, 3171. https://doi.org/10.3390/su11113171

Sanz-Menéndez, Luis, Cruz-Castro, Laura, & Kenedy, Alva. (2013). Time to tenure in Spanish universities: An event history analysis. *PLoS One, 8*(10), e77028. https://doi.org/10.1371/journal.pone.0077028

Schein, Virginia E., Mueller, Ruediger, Lituchy, Terri, & Liu, Jiang. (1996). Think manager – think male: A global phenomenon? *Journal of Organizational Behavior, 17*(1), 33–41. https://doi.org/10.1002/(SICI)1099–1379(199601)17:1<33::AID-JOB727>3.0.CO;2-V

Schiffbaenker, Helene, van den Besselaar, Peter, Holzinger, Florian, Mom, Charlie, & Vinkenburg, Claartje. (2022). Gender bias in peer review panels: 'The elephant in the room'. In Fiona Jenkins, Barbara Hoenig, Susanne Maria Weber, & Andrea Wolffram (Eds.), *Inequalities and the paradigm of excellence in academia*, 109–128. Routledge.

Schmaling, Karen B., & Gallo, Stephen A. (2023). Gender differences in peer-reviewed grant applications, awards, and amounts: A systematic review and meta-analysis. *Research Integrity and Peer Review, 8*(1), 2. https://doi.org/10.1186/s41073-023-00127-3

Scully, Maureen A. (1997). Meritocracy. In R. E. Freeman & P. H. Werhane (Eds.), *Dictionary of business ethics*, 412–414. London: Blackwell Publishers.

Scully, Maureen A. (2002) Confronting errors in the Meritocracy. *Organization, 9*(3), 396–401. https://doi.org/10.1177/135050840293004

Sebo, Paul, & Clair, Carole. (2023). Gender inequalities in citations of articles published in high-impact general medical journals. *Journal of General Internal Medicine, 38*(3), 661–666. https://doi.org/10.1007/s11606-023-07783-4

Sheltzer, Jason M., & Smith, Joan C. (2014). Elite male faculty in the life sciences employ fewer women. *Proceedings of the National Academy of Sciences, 111*(28), 10107–10112. https://doi.org/10.1073/pnas.1403334111

Solga, Heike, Rusconi, Alexandra, & Netz, Nicoli. (2023). Professors' gender biases in assessing applicants for professorships. *European Sociological Review, 39*(6), 841–861. https://doi.org/10.1093/esr/jcac048

Spence, Crawford. (2019). 'Judgement' versus 'metrics' in higher education management. *Higher Education, 77*, 761–775. https://doi.org/10.1007/s10734-018-0303-0

Tauber, Susanne, & Moughalian, Catherine. (2022). Collective system-supporting inaction: A conceptual framework of privilege maintenance. *European Journal of Social Psychology, 52*(7), 1128–1142. https://doi.org/10.1002/ejsp.2761

Teelken, C., Taminiau, Y., & Rosenmoller, C. (2019). Career mobility from associate to full professor in academia: Micropolitical practices and implicit gender stereotypes. *Studies in Higher Education, 46*(4), 836–850. https://doi.org/10.1080/03075079.2019.1651715

Valian, Virginia. (2005). Beyond gender schemas: Improving the advancement of women in academia. *Hypatia, 20*(3), 198–213. https://doi.org/10.1111/j.1527–2001.2005.tb01396.x

Van Arensbergen, Pleun, van Der Weijden, Inge, & van Den Besselaar, Peter. (2014). Different views on scholarly talent: What are the talents we are looking for in science? *Research Evaluation, 23*, 273–284. https://doi.org/10.1093/reseval/rvu017

Van den Brink, Marieke. (2015). The politics of knowledge: The responses to feminist research from academic leaders. *Equality, Diversity and Inclusion, 34*(6), 483–495. https://doi.org/10.1108/EDI-03-2014-0030

Van den Brink, Marieke, & Benschop, Yvonne. (2012a). Gender practices in the construction of academic excellence: Sheep with five legs. *Organisation, 19*(4), 507–524. https://doi.org/10.1177/1350508411418698

Van den Brink, Marieke, & Benschop, Yvonne. (2012b). Slaying the seven-headed dragon: The quest for gender change in academia. *Gender, Work & Organisation, 19*(1), 71–92. https://doi.org/10.1111/j.1468–0432.2011.00594.x

Vinkenburg, Claartje, Ossenkop, Carolin, & Schiffbaenker, Helene. (2022). Selling science: Optimizing the research funding evaluation and decision process. *Equality, Diversity and Inclusion: An International Journal, 41*(9), 1–14. https://doi.org/10.1108/EDI-10-2021-0240

Weisshaar, Katherine. (2017). Publish and perish? An assessment of gender gaps in promotion to tenure in academia. *Social Forces, 96*(2), 529–560. https://doi.org/10.1093/sf/sow062

Wenneras, Christine, & Wold, Agnes. (1997). Nepotism and sexism in peer review. *Nature, 387*, 341–343. https://doi.org/10.1038/387341a0

White, Kate. (2014). *Keeping women in science*. Melbourne University Press.

Whyte, Jamie. (2005). *Crimes against logic*. McGraw-Hill.

Williams, Wendy M., & Ceci, Stephen J. (2015). National hiring experiments reveal faculty preference for women on STEM tenure track. *Proceedings of the National Academy of Sciences, 112*(17), 5360–5365. https://doi.org/10.1073/pnas.1500326112

Winslow, Sarah. (2010). Gender inequality and time allocations among academic faculty. *Gender and Society, 24*(6), 769–793. https://doi.org/10.1177/0891243210382894

Witteman, Holly O., Hendricks, Michael, Straus, Sharon, & Tannenbaum, Cara. (2019, February 09). Are gender gaps due to evaluations of the applicant or the science? A natural experiment at a national funding agency. *Lancet, 393*(10171), 531–540. https://doi.org/10.1016/SO140-6736 (18)32611-4

Wu, Cary. (2024). The gender citation gap. Approaches, explanations, and implications. *Sociology Compass, 18*(2). https://doi.org/10.1111/soc4.13189/

Zhou, Sifan, Chai, Sen, & Freeman, Richard B. (2024). Gender homophily: In-group citation preferences and the gender disadvantage. *Research Policy, 53*(1). https://doi.org/10.1016/j.respol.2023.104895

Zippel, Kathrin S., Ferree, Myra Marx, & Zimmermann, Karin. (2016). Gender equality in German universities: Vernacularising the battle for the best brains. *Gender and Education, 28*(7), 867–885. https://doi.org/10.1080/09540253.2016.1191989

# 2

# TOWARDS INCLUSIVE EXCELLENCE POLICIES AT EUROPEAN LEVEL?

*Maxime Forest*

## Retro-engineering scientific excellence

Scientific excellence emerged as a common goal to be achieved by research-performing organisations – including but not limited to universities – as early as the mid-1980s. Due to the ambivalent relation of this notion to the traditional framework of academic merit, tracking back and summarising the emergence of scientific excellence as the global mantra of the "knowledge-based economy" amounts to a form of reverse (or retro-) engineering. In the glaring absence of a canonical definition of what it is, and of a widely shared historical account of how it emerged, it is indeed necessary to proceed through deductive reasoning, trying to identify the main policy trends and discussions that led research institutions to collectively embrace the pursuit of a certain notion of excellence, and to compete to achieve it.

The conservative (or neoliberal) revolution initiated in the United States under Ronald Reagan's presidency, and during Margaret Thatcher's tenure as Prime Minister in the UK, delegitimated Keynesian principles of sustained, linear growth that had underpinned the continuous expansion of public research institutions and of the access to higher education in liberal democracies after WWII. Following successive economic shocks and consecutive recessions between 1973 and 1987, neoliberal policies aimed at reducing public expenditures and increasing their economic efficiency, while unleashing the Schumpeterian concept of "creative destruction" and promoting groundbreaking innovation by reducing state intervention in the economy.

In the field of research, this global spur in favour of economic *laissez-faire* and greater competitiveness led paradoxically to strengthening the mandate of centralised research agencies (such as the US Science Foundation, founded in 1950), to tighten the access to decreasing public funding by the mean of increased competition, ensured through reinforcing the metrics of scientific excellence.

DOI: 10.4324/9781003564348-5

Following the first academic revolution of the late 19th century, by which scientific research had been incorporated into the social function of universities, the academic world associated with the functioning of traditional universities had established various mechanisms to collegially enforce common quality standards. Depending on domestic cultural and political backgrounds, the emphasis was diversely placed on a spirit of entrepreneurship and individual innovation – as in the US, on the social relevance of science as in European welfare states, notably in Scandinavia – and could incorporate more idiosyncratic characteristics linked to deep-seated cultural norms of excellence, as in Japan (Hisama, 1996). Yet, as neoliberal politics based on the managerialisation of public policies and organisations and the principle of economic efficiency were gaining ground in Western economies, established processes ensuring academic integrity and quality were increasingly criticised for being innovation-adverse, and a narrow set of evaluation devices, such as bibliometrics (measured through citations in high-impact journals), patents, and evidences of research management abilities (notably the one to secure private and public grants), was promoted to support the growing internationalisation of academic production. In many instances, this development was primarily used for shrinking public research, as part of structural, austerity-aimed reforms, as in the case of the "funding decline in the United Kingdom accompanied by the establishment of a governmental evaluation procedure for departments, the closing down of some, the continuance of others – that is, narrowing banks and triage" (Etzkowitz, 1990).

By the mid-1990s, this process had changed the organisation of the academic world, instating centres of excellence that joined public and private research facilities with innovation-driven companies; largely imposing competition-based access to funding over regular, predictable public dotation; popularising university rankings; establishing bibliometrics as the primary instrument to assess individual academic performance; and also measuring scientific impact through market uptake. Deemed an integral part of a second scientific revolution, characterised by a "Triple Helix" of innovation involving research-performing institutions, targeted government funding, and private innovation-focused companies (Leydesdorff & Etzkowitz, 1998; Benner & Sandström, 2000), this multi-layered transformation affecting research funding, selection, competition, and the globalisation of knowledge production was legitimised in the name of a ubiquitous quest for scientific excellence.

## The uneven global diffusion of the excellence paradigm

The steady expansion of this mantra and of its practical consequences for universities and other research-performing organisations did not happen uniformly, nor without raising questions about the validity of this model. In Latin America (Vessuri, 1995) and elsewhere in the Global South, it coincided with structural reform programmes imposed by the IMF and other international funding organisations. In the UK, it preceded the impulse towards a model based on different categories of Higher Education institutions, where a small number would take the dividend of fast internationalisation and research intensiveness, while others would bear the bulk of the massification of higher education. In Northern Europe, this was part of rampant public management policies progressively substituted to the Nordic welfare state model, while in continental Europe – notably in France and Germany – it was at first fended off by strong public research models and, in the latter case, by the upper hand of regional governments over higher education and research.

It is only with the reform of research-funding systems or the creation of new research-funding agencies[1] that competitive, excellence-focused policies hit home through mechanisms such as Labex in France (over 100 labelled initiatives), Scientific Excellence Clusters in Germany (over 70), and the Severo Ochoa and María de Maetzu Centres of Excellence in Spain (around 50). In Central and Eastern Europe, its diffusion coincided both with the second wave of liberalisation after the collapse of state socialism, and the start of the EU integration process by the late 1990s and was largely perceived as a way to break up with ideologically driven research, the inefficiency of state-planned science, and the equalitarian rhetoric linked to the previous regimes, while being met with strong institutional inertia.

As early as 1987, in the first comprehensive attempt to map the origins of scientific excellence as a standard for research and to assess it, Jackson and Rushton (1987) already reunited contributions highlighting elements of continuity between the old meritocratic paradigm and the new one, its uneven consequences across academic fields, and its androcentric character tending to increase, rather than narrow down, women scientists' "competitive disadvantages" rooted into structural gender inequality. And from every country where economic austerity reforms coincided with establishing competitive access to research funding, fostering public–private cooperation and setting up selection and evaluation procedures based on a narrow set of allegedly objective criteria, voices were raised either to encourage local universities and academic structures to embrace the new paradigm (or perish) or more rarely to challenge it, as in Finland, for example (Alestalo, 1993).

The global expansion of the pursuit of scientific excellence thus evidences the crucial interaction of a few global trends (the conservative neoliberal revolution, the rise of public management and its frequent correlation with externally imposed structural reforms programmes, and, after the end of the Cold War, a globalisation process characterised by growing capital and information flows), with domestic patterns in terms of the degree of adherence to the principles of free market competition and of how scientific research was previously structured and funded. Hence, rather than a global concept, it emerged as a myriad of regional and local configurations, where the fundaments of the neoliberal framing of scientific excellence as the driving force of a knowledge-based economy were adopted and enforced differently.[2]

## The EU context: towards a European Research Area

In what had become the European Union in 1992, the diffusion of scientific excellence coincided with the emergence of EU-wide research and innovation policies. These celebrated their 40th anniversary in 2024, as the Framework Programme for Research and Technological Development was launched in 1984, with a budget of 3.8 billion euros. By 1994, two years after the Maastricht Treaty established the European Union, the 4th Framework Programme more than duplicated the budget of the previous one, at 13.8 billion euros, reflecting the ambition of the new supranational superstructure to foster innovation and competitiveness. As the 4th Framework Programme ended, the EU was already on the path to the adoption of the Lisbon Strategy (2000) aimed at enhancing the bloc's competitiveness and establishing itself as a key player of the knowledge-based economy.

Simultaneously, the Bologna process initiated by the Council of Europe in 1998, aimed at mainstreaming European higher education systems through the mutual recognition of degrees and qualifications, streamlining a European Credit System, and establishing quality assurance procedures for Higher Education Institutions. This gradual process, culminated in 2010 with the creation of the European Higher Education Area, comprising 48 states, has increasingly been entangled with the policies pursued in this field by the European Commission. Meanwhile, one of the instruments of the Lisbon strategy was the establishment of a European Research Area, also enshrined in the Treaty on the Functioning of the European Union since the Amsterdam Treaty in 1997, where it is described as an "area in which researchers, scientific knowledge and technology circulate freely, and encouraging it to become more competitive, including in its industry" (Now Art 179 of TFEU). Although processes of certification and selection developed since the mid-1990s were largely aligned with the above-described global trends, the vocabulary of scientific excellence was not widely used until the early 2000s, when research intensiveness and competitiveness were set as core objectives of the EU policy.

This has been the case ever since, notably with the creation of the European Research Council in 2007 as a public body for funding scientific and technological research, composed of an independent Scientific Council and an Executive Agency, in charge of the implementation. The institutionalisation of the ERC was coupled with a considerable increase of the budget allocated to Framework programmes, from 17 to 53 billion, also reflecting two successive EU enlargements, in 2004 and 2007. Though often previously mentioned in Council conclusions and policy papers, and consistent with the quest for greater competitiveness and the adherence to the "triple helix of innovation" evidenced in the Lisbon Strategy, scientific excellence was firmly established as a core objective of EU FPs from the 7th Framework Programme (2007–2013). "Excellent Science" currently constitutes the first pillar of Horizon Europe, the 9th EU FP and is granted alone with 25 billion euros, ascribed to the ERC, the Marie Skłodowska Curie grants, and European Research Infrastructures.

Overall, the notion of scientific excellence has been instrumental in the development of EU research and innovation policies, and the emergence of more unified European Research and European Higher Education areas. Underpinned by principles of selection and competition, measurable performance and merit, and freedom of movement for students and researchers, it has been consistent with the objectives pursued by the European Commission in those realms, and notably with its ambition to support the integration and competitiveness of all EU member states in the global knowledge economy.

## A strange union? Coupling scientific excellence with gender mainstreaming

However, the timeline briefly sketched earlier also reveals an important specificity in the reception of the notion of scientific excellence in the European Union: the dual development followed by its research policies and instruments, and by the institutionalisation of gender mainstreaming as a policy paradigm. Contemporary to the establishment of the European Union, the surge of gender mainstreaming in EU policymaking has followed a similar pace to the one for excellence-driven research and innovation policies. Consolidated in the social pillar of EU policies, launched in 1994, gender mainstreaming first impacted European policymaking through the European Social Fund and through regional funds, the allocation of which has been increasingly conditioned by gender mainstreaming implementation, before spreading towards other policy areas. After the Amsterdam Treaty acknowledged Gender Equality as one of the objectives of the Union in 1997, developments in all EU policy areas were bound to integrate the gender dimension.

The policy sequence leading up to the adoption of the Lisbon strategy (1999–2001) was an important milestone, marked by the first Communication of the European Commission on Women and Science in February 1999, complemented by two follow-up resolutions adopted by the European Council and the European Parliament, while a Women in Science unit was established at the European Commission's DG for Research, and the Helsinki group for women in science was established under the Finnish presidency of the European Council (now Council of the European Union) as an advisory body to the European Commission. Echoing the discussions triggered by the groundbreaking study published in *Nature* by two Swedish physicians, Christine Wennerås and Agnès Wold (1997), and the impulse to gender mainstreaming brought by early adopters – Austria, Finland, and Sweden – joining the EU in 1995, the Commission's communication stressed the need to embed gender equality in the 5th Framework Programme. The next year, the European Technology Assessment Network's (ETAN's) working group on women and science published a report unequivocally titled *Promoting excellence through mainstreaming gender equality*, thus mutually conditioning both objectives (European Commission, 2000). Acknowledging the centrality of the pursuit of scientific excellence in the new EU research and innovation policy, it highlighted the need to promote fairness through access to funding and to modernise peer-reviewing through fighting gender and other biases, further evidencing existing gender gaps in current merit-based science assessment frameworks (Scully, 2002).

In 2001, another resolution of the Council on Science and Society and Women in Science further highlighted the need for the 6th Framework Programme (2002–2008) supporting the creation of the European Research Area to not only address women's participation in science and technology but also consider relevant gender dimensions in research, enjoining the European Commission to collect sex-disaggregated data to monitor women's participation in science. In 2003, a symposium was organised by DG Research at the European University Institute in Florence, devoted to "Minimising gender bias in the definition and measurement of scientific excellence". The ensuing report, titled *Gender and Excellence in the Making* (Husu, 2004), provided a comprehensive panorama of a decade-long critical discussion on the limits of excellence measurement, mentioning the flaws of peer-reviewing as it is practiced, highlighting that "merit and talent are not sufficient conditions to become a successful scientist. Resources, time, social networks, encouragement – unevenly distributed between the sexes – are necessary prerequisites." Rightly observing that bibliometrics are used as "a proxy for excellence, quality, and ability", the report noted that what can be measured is not quality, only representations of quality. Scientists' own reactions to bibliometrics, and the digital revolution enabling the production of more papers with similar efforts, were also identified as the cause of the inflation of scientific production, and aggravating factors for pervasive gender bias, due to the unequal distribution of academic work, women's under-representation in scientific areas privileged by citation indexes, and structural inequalities such as the unequal distribution of unpaid care work. Approaching the pursuit of excellence as "a set of practices framed within a 'discourse' used by the scientific community to organise its self-governance", the contributors to the report consequently challenged a gender-neutral notion of scientific excellence and evidenced that academic practices such as interdisciplinarity or collegial work, more frequently undertaken by female scientists, were also undervalued by the applied metrics of excellence, thus deepening the gender gap.

Although challenging, the report intended to pave the way for a more nuanced and truly objective approach to the measurement of excellence, and established the pursuit of gender equality and the removal of gender bias, as necessary pillars for achieving true excellence. This growing body of evidence, also fed by the She Figures statistical compendium released in 2003 by the European Commission and directly brought to the knowledge of EU policymakers, as well as the lack of progress in women's participation in science and technology following the 5th Framework Programme, led to a significant reframing of the EU's gender equality instruments in its research policy. The 7th Framework Programme was thus characterised by a shift from women to institutions. Rather than supporting women's inclusion by enhancing their "fitness" with established scientific excellence measurement frameworks, it placed a greater emphasis on the responsibility of research-performing organisations, to apply gender-bias-free selection procedures, and to adopt comprehensive policies to tackle gender inequality in academia. One of the instruments made available for this endeavour consists in funding Coordination and Support Action (CSA) projects devoted to advancing gender equality in academia through the design and implementation of comprehensive Gender Equality Plans, and facilitating resources for the ongoing evaluation.

## From fixing women to fixing excellence? Tackling gender bias as a new Horizon

FP7 thus supported a first generation of "sister projects" bound together by the pursuit of a same objective. It also provided additional resources for research-performing organisations to integrate a gender lens in designing and delivering research, in the form of a toolkit and training modules (EC, 2014) or establishing sound criteria for gender training and gender expertise.[3] However, this shift towards transforming institutions did not bring about the expected results. As shown by the low-paced progress registered via She Figures – to date, the only sector-specific gender statistical compendium regularly published by the European Commission – the increasingly sophisticated policy framework established for the ERA was not sufficient to induce significant changes in research-performing organisations, notably regarding how scientific excellence is assessed in practice, leading to reproducing long-identified gender biases. These limitations were further evidenced in a new wave of reports, such as *Structural Change in Research Institutions: Enhancing Excellence, Gender Equality and Efficiency in Research and Innovation* (EC, Sanchez de Madariaga, 2012). Pinpointing the notion of structural change (as opposed to adapting women to the current functioning of the research system), the report was taking stock of the progresses made in establishing a robust policy framework, but highlighted the persistence of opaque decision-making procedures, adverse institutional practices, and unconscious bias skewing the assessment of excellence. Noting that "what characterizes excellence is generally not itself subjected to scientific evaluation", the report advocated critically approaching this socially constructed concept, considering context, culture, and subtle biases. Not much different in its diagnosis from the previous wave of reports a decade earlier, it brought about recommendations to bridge the gap between the alleged pursuit of excellence and achieving gender equality in research and academia.

Roughly a decade after the launch of EU women in science policies, the shift to Horizon 2020 in 2014, a new Framework Programme granted with 77 billion euros, reflected this critical assessment. New criteria and indicators were introduced for project selection, regarding the integration of the gender dimension in science (EC, 2013, 2015), and additional resources were brought to support structural change policies, notably through funding a Gender Equality Academy, aimed at providing quality, tailor-made capacity-building to research-performing organisations, or UniSAFE devoted to tackling sexual and gender-based violence in academia, notably through the first EU-wide survey on the issue. The European

Institute for Gender Equality, founded in 2010, also introduced two major innovations: a Gender Equality Index, computing data in various fields including access to education (2013) and in 2016, the first version of the GEAR tool (standing for Gender Equality in Academia and Research),[4] offering step-by-step guidance for universities and other research organisations to implement holistic gender equality plans. Meanwhile, the threshold was placed higher for EU-funded GEP projects, encouraged to also involve research-funding bodies, due to their excellence standards-setting role.

Horizon 2020 represented a qualitative step forward and the sustained attention of the European Commission for a gender-sensitive notion of scientific excellence simultaneously contributed to significant developments in the legislative and policy frameworks of EU member states. Following pioneers of gender mainstreaming in science such as Finland or Spain, countries including France, Germany, Ireland, and later Portugal and Belgium, adopted new provisions to support Gender Equality Plans in universities also from the national and regional governance levels. By 2016, about half of EU member states had already adopted such provisions (EIGE).

Mid-term of Horizon 2020, DG Research commissioned an interim evaluation of the new criteria on gender equality and the gender dimension (de Cheveigné et al., 2017), which evidenced the persistence of a gender-blind notion of scientific excellence, since a majority of the projects scrutinised did not consider the gender dimension in most disciplinary areas (Figure 2.1) and encompassed little or no

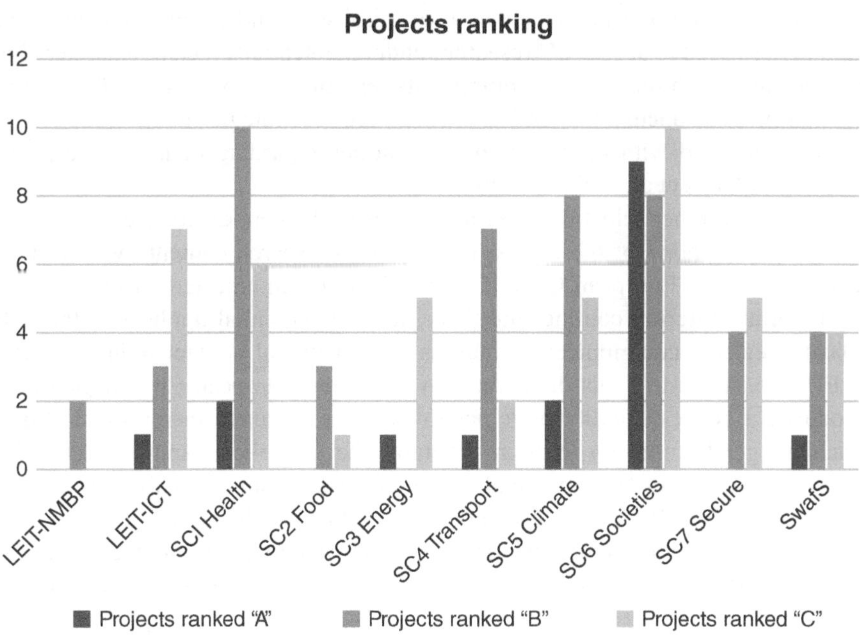


**FIGURE 2.1** Ranking of projects per disciplinary EU calls, based on gender criteria[5]

gender expertise. Scholarly contributions, as those by Van den Brink and Benschop (2012a, 2012b) on gender practices in the construction of academic excellence (depicted as the quest for the five-leg sheep) and institutional resistances to structural change policies in the academia (amounting to "slaying the seven-head dragons"), vividly depicted the enduring gendered notion of academic excellence.

From the mid-2010s, several contributions have focused on the micropolitics of gender equality and excellence at university level, highlighting, for instance, discursive practices legitimising gender equality policies in the name of excellence and performance (see O'Connor, this volume; Clavero & Galligan, 2021; Linková & Mergaert, 2021). Also further deepening the understanding of gender bias in peer-reviewing brought by seminal works of Wennerås and Wold (Peterson & Husu, 2023), this relatively abundant literature stresses the gap between policy intentions and reality, and analyses how research-performing and research-funding bodies passively and actively resist a gender-sensitive notion of merit and excellence through (gendered) social reproduction, self-legitimation, and inertia.

## Going stronger, wider, and intersectional: a new paradigm for Horizon Europe

This cumulative effort, jointly undertaken by gender scholars, experts, and policy networks and competent units and agencies of the European Commission, culminated in the latest (9th) Framework Programme, Horizon Europe, granted 95.5 billion euros and launched in 2021. As of January 2022, Horizon Europe introduced the adoption and implementation of a holistic gender equality plan as an eligibility criterion to access EU research funding. Potentially a game-changer, this criterion came with mandatory requirements regarding the process leading to the adoption of such documents, and to their nature of strategic, publicly available commitments, and with precise recommendations regarding their coverage and scope (see, Matias et al. in this volume).

Additionally, other selection criteria, in terms of the gender balance of research teams and integration of the gender dimension in research content, were further enhanced, and the European Commission actively communicated, prior and after the eligibility criterion coming into force, towards potential applicants, through dedicated events, communication material, and national contact points (NCPs) for EU FPs. As part of this dissemination effort, the European Commission also introduced a gender equality prize acknowledging sustained institutional efforts to achieve gender equality through structural change policies and mandated the European Institute for Gender Equality to undertake a comprehensive update and upgrade of its GEAR tool. Another notable development is the new emphasis on the intersectional dimension, which is supported by the recently adopted work programme for Horizon Europe and the EU Strategy for Equality of LGBTQ People (2020–2025). Universities and research bodies are thus encouraged to implement "inclusive" equality plans, and the Commission has funded the creation of a "European Centre of Excellence for Inclusive Gender Equality in Research and Innovation" in 2022.

This stronger emphasis on institutional capacity was also mirrored at the level of member states, with the adoption of the Ljubljana declaration under the auspices of the Slovenian presidency of the Council of the European Union, in November 2021. Acknowledging gender mainstreaming as a horizontal principle and gender equality, together with inclusiveness, to contribute to research integrity and social responsibility, the Declaration communicates a series of commitments for the member states to strengthen their policy framework in support of inclusive gender equality plans. Highlighting the persisting gender inequalities revealed by the COVID-19 crisis, it also advocates for synergies between the EU research policy, the European Recovery Facility, and the European Green Deal, arguing for a strong correlation between achieving gender equality and inclusion and meeting technological or environmental innovation challenges.

Another lesson drawn from the implementation of previous EU FPs was the remaining gap among member states, in terms of their adherence and policy commitment to a gender-sensitive notion of academic excellence. This gap largely follows the division between higher and lower research-intensive countries (in terms of the share of their GDP devoted to R&D and of their ability to draw from EU resources), with the latter mostly located in Central and South-Eastern Europe. Rooted in different understandings and experiences of gender equality, and path-dependent features also affecting the domestic organisation of research and academic institutions (Kwiek, 2012; Antonowicz et al., 2017) – particularly with regard to the experience of State socialism – this gap is double, binding together lower research and innovation intensiveness with lower gender equality scores as shown in EIGE's index.[6] This conjunction has been effectively mobilised at the EU level, to encourage the so-called "widening countries", to embrace both agendas at once. Hence, resources have been redirected to those countries through the WIDERA calls, which intend to widen EU policy for gender equality both geographically and thematically, by adopting an intersectional lens.

To date, both intentions remain largely unfulfilled. In terms of bridging the gender and innovation gap in Central-Eastern Europe, current WIDERA projects tend to indicate that long-documented East-West differences persist with regard to the Europeanisation of gender equality policies (Lombardo & Forest, 2012; Forest, 2021), in a context of growing polarisation over gender issues (Kuhar & Paternotte, 2017), as reflected by successive attempts to enforce gender studies bans in universities (as in Hungary, Romania, and Bulgaria) (Ergas et al., 2022). Such differences also prevail in the lower standards being maintained for gender equality policies in research-performing organisations of the region. In that sense, the pioneering developments at the University of Lodz as part of RESET are indicative that inclusive gender equality plans can nonetheless be brought forward in adverse contexts and in anticipation of domestic policy change.

In terms of adopting an intersectional approach, the lack of evidence of effectively implemented intersectionality even in those areas where the concept has long been articulated, such as international development (Forest & Foreste, 2023), deprives academic institutions of valid examples from which to draw inspiration. This agenda might also be complicated by the fact that unlike gender mainstreaming, intersectionality is all but a "concept-method" (Jacquot, 2010), and that any sort of diversity mainstreaming addressing various discrimination grounds, falls short of a truly intersectional approach, the purpose of which is to tackle the dynamic relation among those grounds, rather than their mere addition (Hancock, 2007).

## Conclusions

Presumed to be meritocratic, scientific research has long required tools to assess the merit of individuals and the performance of institutions. However, this old meritocratic framework, which differed across disciplines, countries, and institutions, was largely subsumed, from the 1980s, into the competitive pursuit of scientific excellence, on the basis of free-market principles and the alleged benefits of a triple helix linking governments, research organisations, and innovation-driven companies. Through the creation of an academic market and the transnational circulation of knowledge, and propelled by the diffusion of the neoliberal principles of public management, what has been described as the "second academic revolution" progressively transformed the way research and academic systems operate. Evaluation and reporting tools have proliferated, in the name of economic efficiency, research quality, and transparency.

Nearly as early as the notion was being popularised, critiques emerged, to challenge the pretension to objectively measure scientific excellence through a handful of quantitative instruments. Gender scholars and non-gender scholars alike have brought evidence that far from removing bias from science evaluation, those instruments were reproducing, industrialising, and "objectivising" them. Much debated contributions on the negative impact of peer-reviewing or individual competition on women's career prospects and recognition have constituted a dynamic and stubbornly critical field of study and fed initiatives on both sides of the Atlantic to mitigate the collateral damages of scientific excellence. Beyond gender, this notion has also been criticised for the geographical or disciplinary inequalities it creates (Musselin, 2021), or the way it reproduces post-colonial hierarchies in knowledge production and circulation.

However, it is at the EU level that the acculturation to an inclusive approach to excellence, focused on preventing biases affecting the selection of academic workers and the innovations they produce, seems to have taken hold. This highlights the paradox of an environment – the European Research Area – built around the promotion of excellence, but whose governance mechanisms (funding, selection criteria, and programmes), long framed as the trojan horses of neoliberalism, are now being used to promote a more inclusive approach to scientific innovation.

Although the limitations of this dual agenda are well-documented, each EU framework programme has brought innovations that have cumulatively strengthened the incentives for research-funding and research-performing organisations in the EU and associate countries to achieve inclusive scientific excellence and lately, to do so by adopting an intersectional lens. Whether this increasingly consistent policy framing will ultimately be another example of the co-optation of gender mainstreaming by a neoliberal agenda, or the instrument of a paradigm shift in line with the principles of responsible research and innovation by and for society, remains to be demonstrated.

## Notes

1 As the French ANR in 2005 or the Spanish AEI in 2016.
2 In this respect, the notion of excellence was not unrelated to two interrelated objectives: increasing transparency in the operation of the academic world, and combating mandarinism and "localism", consubstantial phenomena based on co-optation and nepotism.
3 This was the objective of the OPERA stream of the QUING project, funded under FP6 (2007–2011).
4 https://eige.europa.eu/gender-mainstreaming/toolkits/gear?language_content_entity=en.
5 de Cheveigné et al., 2017, p. 25. A>C= higher>lower consideration of the gender criteria in submitted proposals.
6 https://eige.europa.eu/gender-equality-index/2023.

## References


Alestalo, Marja. (1993). The rise of neo-liberalism in Finland: From the politics of equal opportunity to the search for scientific excellence. *Science Studies, 6*(2), 35–47.

Antonowicz, Dominik, Kohoutek, Jan, & Hladchenko, Myroslava. (2017). The roads of excellence in Central and Eastern Europe. *European Educational Research Journal, 16*(5).

Benner, Mats, & Sandström, Ulf. (2000). Institutionalizing the triple helix: Research funding and norms in the academic system. *Research Policy, 29*, 291–301.

Clavero, Sara, & Galligan, Yvonne. (2021). Delivering gender justice in academia through gender equality plans? Normative and practical challenges. *Gender, Work & Organization, 28*(3), 1115–1132.

de Cheveigné, Suzanne, Knoll, Bente, Bustelo, Maria, Engebretsen, Eivind, & Sandström, Ulf. (2017). *Interim evaluation: Gender equality as a crosscutting issue in Horizon 2020. European Commission*. Luxembourg: Publications Office of the European Union.

Ergas, Yasmine, Kochkorova, Jazgul, Pető, Andrea, & Trujillo, Natalia. (2022). Disputing "gender" in academia: Illiberalism and the politics of knowledge in politics and governance. *Politics and Governance, 10*(4), Gender and Illiberalism in Central Europe (Special Issue).

Etzkowitz, Henry. (1990). The second academic revolution: The role of the research university in economic development. In S. E. Cozzens, P. Healey, A. Rip, & J. Ziman (Eds.), *The research system in transition, NATO ASI series* (Vol. 57). Springer.

European Commission. (2000). *Science policies in the European Union – promoting excellence through mainstreaming gender equality: A report from the ETAN expert working group on women and science*. Luxembourg: Publications Office of the European Union.

European Commission. (2013). *Gendered innovations: How gender analysis contributes to research*. Luxembourg: Publications Office of the European Union.

European Commission. (2014). Gender in EU funded research and innovation: Toolkit. Luxembourg: Publications Office of the European Union.

European Commission. (2015). *Indicators for promoting and monitoring responsible research and innovation: Report from the expert group on policy indicators for responsible research and innovation*. Luxembourg: Publications Office of the European Union.

Forest, Maxime. (2021). Europeanization. In G. Abels, A. Kriszan, H. Mac Rae, & A. Van der Vleuten (Eds.), *The Routledge handbook of gender and EU politics,* 30–42. Routledge.

Forest, Maxime, & Foreste, Christlord. (2023). *L'approche intersectionnelle: Généalogie, contestations et perspectives pour le champ du développement international*. Paris: Agence Française de Développement.

Hancock, Ange-Marie. (2007). When multiplication doesn't equal quick addition: Examining intersectionality as a research paradigm. *Perspectives on Politics*, *5*(1), 63–79.

Hisama, Kay Keiko. (1996). Excellence in scientific research and writing. *Bulletin of Allied Medical Sciences Kobe, 11*, 179–184.

Husu, Lisa. (2004). Gate-keeping, gender equality and scientific excellence. In *European Commission: Gender and excellence in the making*, 69–76. Brussels: Directorate-General for Research and Innovation.

Jackson, Douglas J., & Rushton, Phillippe. (Eds.). (1987). *Scientific excellence: Origins and assessment*. Seven Oaks: Sage Publishers.

Jacquot, Sophie. (2010). The paradox of gender mainstreaming: Unanticipated effects of new modes of governance in the gender equality domain. *West European Politics, 33*(1), 118–135.

Kuhar, Roman, & Paternotte, Davod. (2017). *Anti-gender campaigns in Europe*. London: Rowman & Littlefield.

Kwiek, Marek. (2012). Universities and knowledge production in Central Europe. *European Educational Research Journal, 11*(1).

Leydesdorff, Loet, & Etzkowitz, Henry. (1998). The triple helix as a model for innovation studies. *Science and Public Policy, 25*(3), 195–203.

Linková, Marcela, & Mergaert, Lut. (2021). Research policy. In G. Abels, A. Kriszan, H. Mac Rae, & A. Van der Vleuten (Eds.), *The Routledge handbook of gender and EU politics*, 314–326. Routledge.

Lombardo, Emmanuela, & Forest, Maxime. (2012). *The Europeanization of gender equality policies: A discursive and sociological approach, Gender and politics series*. Palgrave Macmillan.

Musselin, Christine. (2021). University governance in meso and macro perspective. *Annual Review of Sociology, 47*, 305–345.

Peterson, Helen, & Husu, Lisa. (2023). Online panel work through a gender lens: Implications of digital peer review meetings. *Science and Public Policy, 50*(3), 371–381.

Sanchez de Madariaga, Inès. (Rap.). (2012). *Structural change in research institutions: Enhancing excellence, gender equality and efficiency in research and innovation. European Commission*. Luxembourg: Publications Office of the European Union.

Scully, Maureen A. (2002). Confronting errors in the meritocracy. *Organization, 9*(3), 396–401.

Van den Brink, Marieke, & Benschop, Yvonne. (2012a). Gender practices in the construction of academic excellence: Sheep with five legs. *Organization, 19*(4), 507–524.

Van den Brink, Marieke, & Benschop, Yvonne. (2012b). Slaying the seven-headed dragon: The quest for gender change in academia. *Gender, Work & Organization, 19*(1), 71–92.

Vessuri, Hebe. (1995). Recent strategies for adding value to scientific journals in America. *Scientometrics, 34*(1), 139–161.

Wennerås, Christine, & Wold, Agnès. (1997). Nepotism and sexism in peer-review. *Nature, 387*, 341–343.

# 3

# SCIENTIFIC EXCELLENCE AND EQUALITY

## The perspectives of university personnel in four European countries (Greece, France, Poland, and Portugal)

*Marion Paoletti*

### Introduction

Over the past 20 years, scientific excellence has become a watchword for institutions and staff alike, who are now engaged in competition on local, national, European, and global scales. At institutional level, international rankings act as yardsticks. At the level of academic careers, the measurement of "performance" is based primarily on three criteria: quantity of publications, success in calls for projects, and internationalisation. Excellence as an imperative for staff and institutions has been the subject of considerable debate for several years now, driven in particular by feminist criticism. The criteria for excellence, supposedly neutral and objective, in fact appear to favour men, calling into question the notion of merit embedded in that of excellence (see Pat O'Connor's chapter in this book and the general introduction).

Does this debate between equality and excellence exist at the level of European university personnel? What meaning do these employees give to the notion of scientific excellence, which is at once central and vague? Confronted with persistent career inequalities between men and women, are they questioning the neutrality of the notion? And if so, in what ways?

#### *Interplay between the themes of excellence and equality*

Julia Nentwich and her co-authors have already analysed three types of encounter between discourses on excellence and those on equality on the basis of 12 interviews with national experts on higher education in Switzerland (Nentwich et al., 2019). In the first type of discourse, excellence and equal opportunity are seen as contradictory, with meritocracy appearing as a matter of course. In the second

DOI: 10.4324/9781003564348-6

encounter, the relationship between excellence and equal opportunity is problematised, with the model of excellence appearing problematic for women. In the third encounter, excellence is seen as highly dependent on equal opportunity, justifying institutional change.

To what extent are these three types of discourse found among academic staff? Do they believe in the virtues of academic excellence to distinguish the best, and are they attached to the status quo without intervention in favour of equal opportunity? On the contrary, are the notion and its criteria criticised in the name of gender and diversity inequalities in recruitment and careers? To what extent do these criticisms call into question the idea that equality policies are a prerequisite for the credibility of excellence? And if so, among which type of personnel? Compared with the interviews that Nentwich and her co-authors conducted with Swiss university policy experts in the mid-2010s, is there any change in the perceptions of workers in European university organisations in the early 2020s?

To answer these questions, between May and October 2021, we organised 15 focus groups (FG) in the four project universities implementing gender equality plans: Aristotle University of Thessaloniki (AUTh), University of Bordeaux (UBx), University of Lodz (UL), and University of Porto (U.Porto). The interview script for the FG was built around several topics: scientific excellence, vertical segregation and leadership, family-friendly policies, gender dimension in research and knowledge transfer, gender biases and stereotypes, sexism and sexual harassment, and gender equality measures, where we aimed to gather suggestions and understand the attitudes towards different measures.

All of these FGs began with a section devoted to scientific excellence, with an initial general instruction: "For you, what do the term scientific excellence mean?". It was only at the end of the first phase of discussion that the moderator introduced, if necessary, the question of neutrality. "Do you think that gender inequalities in academic careers are compatible with scientific excellence?" The dynamics of discussion were then left as free as possible, in order to see how the themes of equality and excellence interacted, and to what extent equality issues did or did not call into question excellence as perceived by academic workers. This chapter therefore focuses on this introductory stage of the FG, firstly in terms of the general perception of scientific excellence, and secondly in terms of any debate on the relationship between equality and excellence.

### *A few methodological remarks*

The 15 FGs organised at the four universities were set up on a homogenous and professional basis. These groups are both "natural" (produced by a shared professional affiliation) and artificial (created by and for the occasion). The target groups were teachers, researchers, university staff members, and top management, with the greatest possible variety within each category. The way these choices were put into practice depended on the institutional context, with local variations.

**TABLE 3.1** Distribution of participants, and modality of the RESET Focus Group (2021)

| | *AUTh* | *UBx* | *UL* | *U.Porto* |
|---|---|---|---|---|
| Administrative Staff (AS) | 2 female<br>1 male | 11 female | 5 female | 7 female<br>3 male |
| Top Management (TP) | – | 6 female<br>3 male | 3 female<br>1 male | – |
| Teachers (only) (T) | 1 female<br>1 male | – | – | 9 female<br>2 male |
| Teachers-Researchers (TR) | 2 female<br>2 male | 6 female<br>2 male | 5 female<br>1 male | – |
| Researchers (R) | 1 female<br>1 male | – | – | 3 female<br>1 male |
| Deans and Vice Deans (DVD) | – | – | 3 female<br>4 male | 8 female<br>1 male |
| Modality | Online | In presence | Online | Online |

29 administrative directors, 37 teachers and/or researchers (all categories), 13 members of top management (mainly vice presidents at university level), and 16 deans or vice deans (at faculty level) took part in the focus groups.

By being as close as possible to employees, we can hope to gather their views on the themes of scientific excellence and equality, but they cannot of course be isolated from their institutional and national affiliations. Comparisons are difficult because there are a variety of variables to take into account. The institutionalisation of equality policies at the national level and the existence of a sectoral policy at the academic level vary in France, Portugal, Greece, and Poland. Employees at the four universities are not exposed to the same policy of excellence on the part of their university and their country. On the one hand, the word has become widely disseminated among UBx workers 15 years after the launch of a national policy of creating sites of excellence through a process of selection and concentration of public funds (from which UBx benefited). It is described as "the great university race" (Musselin, 2017). On the other hand, its rise to prominence seems recent and significant at UL and widely shared at U.Porto, which dedicates a page of its website to "U.Porto in the Rankings". The emphasis and policies on excellence vary in intensity from one institution to another, which explains, for example, why UL or AUTh administrative staff do not have much to say about scientific excellence, unlike their French or Portuguese counterparts. In addition, while the four universities share some common features when it comes to equality (a high proportion of women in administrative positions, fewer in scientific jobs, and a tendency for women to disappear from Management and decision-making), the situation is mixed on certain points. The feminisation of scientific fields is more advanced at UL and AUTh; the group of decision-makers is very feminised within UL, which tends to endorse the idea that "equality is already there" and not to perceive equality as a problem at AUTh or UL, at least in relation to administrative staff. On the other hand, at UBx and U.Porto, all the staff are seizing the opportunity to discuss scientific excellence.

Among FGs' participants, the imbalance to the detriment of men is clear: of the almost one hundred people who took part in the discussions, more than three-quarters were women. This is true for the highly feminised administration sector and even more remarkable for the less feminised research sector. The groups set up by AUTh had almost equal numbers of women and men, but they also had the fewest participants.

This selection bias must be borne in mind when interpreting the results. The participants clearly had in mind to take part in a project designed to promote gender equality within their institution, which may have influenced the identity and motivations of the volunteers as well as the type of discourse that could be expressed in this type of arena. A call for volunteers, targeted at each professional group, was sent to all employees, explicitly inviting them to come and "exchange and compare experiences and points of view on the subject of gender equality, and thus identify the main obstacles and levers for gender equality in higher education and research". The term "scientific excellence" did not appear in the invitation. The decisive moment of recruitment proved to be difficult, with a low willingness to participate on the part of staff, particularly men, which required the intervention of vice presidents, for example, to motivate administrative managers. This intervention was effective, with 29 administrative managers ultimately participating (including only 4 men).

Once again, the focus group method proved to be a successful way of generating personal viewpoints on a theme, even though only the focus groups at UBx could be conducted face-to-face (due to COVID). The transcripts were first written in local languages before being translated. Special efforts were made to be as accurate as possible in terms of what happened in the sessions. The analysis is carried out ex post using a reading grid, built around several questions, to categorise the recurring themes and their relationships. When asked to define the meaning of scientific excellence, for their university and for themselves, did the participants develop a critical or supportive approach? What are the foundations (particularly ethical) of their position? Do they spontaneously address the issue of inequality, or not? When the moderator introduces this dimension, do they reject or develop expand upon it? And if excellence is criticised in the name of effective equality of opportunity, are any implications developed? Analysing individual opinions essentially involves observing who says what, while paying attention to the dynamics of the discussion.

## *Chapter outline*

Beyond the differences in context that shape the discourse, criticism of scientific excellence is considerable and can be found in each of the universities, particularly among teachers and/or researchers. According to the results obtained by Julia Nentwich and her co-authors (admittedly based on interviews with experts, not academic workers), the criticism of scientific excellence is severe and shared. Ambivalent, the notion appears as an ideal in perdition (1). The criticism does not necessarily lead to the notion's abandonment: a minority of participants consider that it remains the guarantor of equality and merit. But the vast majority of participants explicitly identify the androcentric biases behind the ideal type of excellent researcher (hyper-productive, internationalised, with a fast, linear career). This finding may (but not necessarily) lead to a new imperative for the credibility of scientific excellence: the need to actively promote institutional change in favour of equality and diversity in academic careers (2).

## An ideal in perdition? Ambivalence of the notion

The notion of excellence gives rise to considerable ambivalence on the part of individuals and groups who discuss its meaning. A priori positive, conveying moral values such as high standards, investment, and rigour, which make it possible to rank researchers and institutions, excellence tends to become, particularly for researchers, a "fashionable" word that is "overused", and whose meaning is considered lost: if everyone has to be excellent, what is the meaning of the distinction conveyed by the notion? The pressure, particularly on scientific productivity, and the resulting acceleration, clash with a definition of excellence based on professional values.

### *Surpassing yourself or exceeding others: moral qualities and results*

On the positive side, excellence refers to moral attitudes and the means used rather than measurable results, enabling people to surpass themselves. At work, "for me it's perhaps more a question of giving the best of oneself" (UBx, AS) but also in one's personal life: "excellence means trying to improve every day, in one's life" (U.Porto, AS). The notion is associated with "being demanding in the sense of being rigorous" (U.Porto, Deans/Vice Deans – DVD). It means, for example, "setting objectives and surpassing them" (U.Porto, DVD).

The concept seems close to that of meritocracy or elitism in that it singles out a remarkable minority and establishes a hierarchy of researchers and institutions. The emphasis is then not so much on resources as on results: "It's a word that's supposed to evoke what goes beyond a group. Some things are good, and there's one thing that's excellent, and that's what goes beyond" (UBx, TM). "For us, excellence is a measure. In other words, things happen, someone measures them, then there is a distribution and the excellent ones are those in the top 5" (U.Porto, R.).

The process of distinction and hierarchisation is also at work for institutions. Although European or world rankings are mentioned – "excellence also depends on rankings in international rankings" (UBx, AS) – the reference framework remains national.

> In any case, there are benchmarks that allow us all to consider that our university is an excellent university in terms of more objective indicators: demand indicators, results indicators, and comparative indicators, with figures, even though we are tired of putting them in reports, and compared to others, we always have higher figures, and that allows us to talk about excellence.
>
> *(U.Porto, DVD)*

"Whenever I hear the expression 'university of excellence', I always think of the objective criteria at the scientific level. (. . .) These are all indicators that we can compare with other universities and say, 'this university has better indicators than the others'". (U.Porto, R.)

Therefore, it seems, from the participant's views, that a university's excellence is measured by its "attractiveness", particularly to students.

### *"Expectations that sometimes exceed what we are capable of doing"*

Scientific excellence is seen as a notion that should be positive, but it has lost its meaning because it is "too much in use" and "too fashionable". It is perceived as being used to put (unnecessary) pressure on research workers.

> I think it's a notion which is objectively rather noble, to strive for excellence is still, um . . . It starts from a good feeling, I'd say, but at the same time, it puts constant pressure on you, because it means that there are expectations to reach a level that must always be excellent, that is, but that must continue to be so all the time, I find that it goes with expectations that sometimes go beyond what we're capable of doing, of managing, so here I'd say that there's a positive point and then all the difficulty that goes with it.
>
> *(UBx, TR)*

> For me, excellence is a buzzword! It's a word that's already so overused, that's used in so many different contexts . . . and with . . . sometimes even as a way of putting pressure on people – who, I think, are generally already highly motivated, and who don't often need this pressure – to go beyond . . . which, I think, is sometimes counter-productive. Because it's too much, isn't it? It ends up making people's lives less . . . less balanced, let's say. It's a word I don't like personally, even though I liked it a lot a few years ago. It's a word that made a lot of sense to me a few years ago when I started my academic career.
>
> *(U.Porto, R.)*

The criticism mainly concerns the acceleration of time:

> It's a bit strange for me, but I associate it with the notion of acceleration. In other words, in our work, we're constantly rushing through Initiatives of Excellence and calls for projects, all the time, without taking a step back, this kind of acceleration of time. It's a form of pressure too. So, I agree that it should be positive, but it's true that over time . . . We haven't said it, but for me, it's quite true that there's an aspect of speed that exists with this label of excellence.
>
> *(UBx, TM)*

### *A conflict of definition? Managerial expectations and professional ethos*

The agents in charge of the excellence policy have clear expectations of the projects funded by the EU at universities. As one of them explains, in 2007, excellence referred to the state of the art, the soundness of the methodology, and the relevance of the objectives. Since 2014, interdisciplinarity, the gender dimension, and the use of stakeholder knowledge have been added. Since 2020 and Horizon Europe, the participatory dimension has been strengthened by including open science practices, sharing and management of research output, and engagement of citizens, users, and impact. The same person notes that the ability to integrate all of these criteria, which are widely used by national research-funding agencies, leads to a division between researchers:

> The researchers who manage to keep up with the times are those who already have one foot in the game and who sometimes play the game just to play the game. They do it just because they have to meet an additional criterion.
>
> *(UBx, AS)*

The criteria used to assess the content of a project can lead to saturation among researchers. It is now more a question of meeting the criteria than of doing good science, with the notion of excellence appearing as a "whip" to put pressure on researchers.

> And then you start using excellence, not as a qualification, but everyone starts working for it, and then you put another measure in here, and at some point, we're all working for everything and we're in the situation we're in now, and that's what I think my colleagues have said . . . . excellence used as . . . it's not even a carrot, it's almost a whip, isn't it? To put pressure on people who, I think, are already very motivated in general.
>
> *(U.Porto, R.)*

A tension arises between "pure scientific excellence and excellence measured by the university administration" (UBx, TR). A certain contradiction between managerial expectations of excellence and internal professional expectations is apparent.

While most staff, especially researchers, define scientific excellence in terms of the external expectations placed on them, a minority, especially UL researchers (who are also vice deans), define excellence as a concept specific to the research profession. They refer to the ethics of scientific research, which presupposes a work that is well done, based on the implementation of a rigorous methodology that complies with the rules of the art. Therefore, they refer to the notion of scientific integrity and the ethics that are specific to the profession and require time.

Criticism of the limits of bibliometrics quantification for the quality of research is along the same lines.

> It's a word that really annoys me. Because I find that it's used a lot for researchers, for example, creating exactly the opposite . . . Excellence is perceived as publication, publication, publication . . . and publication, publication, publication is not always excellent work. In my field at least, you often have to stop and think, to introspect, to think more deeply, which is often not possible when you're in a hurry to publish. You end up doing a lot of work, but perhaps not excellent. So, unfortunately, excellence ends up having a rather negative connotation from what I see around me.
>
> *(U.Porto, R.)*

> We also hear about the race to publish, the need to publish a lot, but there's also the perception that there are other types of excellence that could be recognised: the idea that it might actually be more interesting to produce in-depth work, even if it means not cutting it into 50 slices . . . Taking one's time is necessary, but it's the 'anti-model' in a way, compared to the quantitative notion of excellence, which everyone knows has all kinds of biases and flaws.
>
> *(UBx, TR)*

"Fashionable themes" in calls for projects are also highlighted:

> So I agree, it can be quite fashionable, and if we work on this theme, it's true that we can showcase our excellence, but if we don't work on this theme at all, then all our work is relegated to second-class status.
>
> *(UBx, TR)*

Several teachers and/or researchers complained about the restrictive definition of scientific excellence, reduced to publications, and insisted on the time devoted to teaching and pedagogical innovation, which is insufficiently taken into account.

Finally, difficulties in recruitment and promotion in the academic sector were spontaneously mentioned as limiting the relevance of the imperative of excellence. The precariousness of researchers and the difficulty of obtaining a permanent position and then progressing through the various career stages were associated with a form of "hypocrisy" contained in the notion of excellence. "The precariousness in which we live, in which researchers live . . . At this level, there is no excellence" (U.Porto, R.). This social aspect was raised particularly by AUTh's group of researchers. The lack of economic support for young researchers means that they have to work in parallel with their studies and their entry into an academic career, which makes the ideal of excellence unattainable. As one U.Porto researcher put it, "There's no point in being the best in science or having more publications if the result is that people feel unhappy in their work, that they don't have good working conditions." This tension is reflected later in the transition to the upper echelons of the scientific career. Someone mentioned the "huge gap" between the omnipresence of the word excellence and the scarcity of recruitment and promotion, a form of "hypocrisy" according to her. "It's a bit like queuing at the butcher's, you have a ticket, and you wait until you're um . . . at the right moment. So, I find that something that's complicated to live with, quite hypocritical" (UBx, TR). Excellence, which is supposed to translate at an individual level into a stable position and career progression, comes up against the low number of possibilities and thus fuels scepticism.

Restrictive, pressurising, acting like a "whip", and no longer compatible with the time required for quality research, the notion of scientific excellence seems to have lost its appeal in recent years. It is criticised internally, spontaneously, and independently of gender inequalities, whereas socio-economic conditions are mentioned in all four universities.

## Links in the discourse on excellence and equality

The theme of gender inequality associated with excellence does not arise spontaneously in most cases, except for UBx (for all groups). It is only when they are explicitly asked (for example, by mentioning the leaky pipeline phenomenon in scientific careers) that staff agree to reflect on the relationship between gender and excellence. Three types of discourse then emerge: a discourse mentioned by fewer participants according to which the criteria of excellence are neutral and guarantee objectivity. A discourse held by the majority that describes scientific excellence as androcentric, signaling a circulation and appropriation of feminist work on the gender of academic professions compared to the results of Nentwich et al. Finally, as a result of the discussion process, the idea emerged that effective equality of opportunity is a condition for saving excellence as a concept close to merit.

### *Excellence versus equality*

The people most committed to defending the neutrality and objectivity of the current criteria for excellence are three men (R or T-R). They suggest that a gender-blind approach is necessary and avoid considering a gender perspective within the academic organisation and community. Social differences between women and men, while they have consequences for women's academic performance, do not imply that the notion of scientific excellence should be called into question. Social differences between women and men, while they have consequences for women's academic performance, do not imply that the notion of scientific excellence should be called into question. Some teachers and researchers believe that scientific excellence is not directly linked to gender discrimination and it does not trigger unequal treatment in research. They rather point to social inequalities, schematic gender roles, and imbalances between work and life duties as reasons for bias in the scientific sphere. Scientific excellence is, therefore, a gender-blind concept, not perceived as problematic or unjust.

In the same vein, another researcher suggests that blind evaluation procedures (for articles or projects) are the best guarantee of "objective" equality and the preservation of excellence. Another one considers that researchers should be classified as "good" or "bad" employees and that their membership in one or other of these categories should be determined by their academic value and professional efficiency.

As feminist literature on the topic suggests (Farinaz, 2015; Ferree & Zippel, 2015), in this vision, excellence is used to mask inequality, and meritocracy is taken for granted. The persistence of this belief can be seen through indirect evidence of the de-legitimisation of positive action to encourage the local promotion of women to professorships. As merit is defined in terms of individual success, positive action programmes are perceived as illegitimate because they provide resources to which only certain groups have access (Augoustinos, 2005). A female researcher at UBx reported that despite the very low presence of female teachers in mathematics, for example, colleagues refused to promote them locally, arguing that "When we promote people, we don't promote them locally because . . . we're really national, excellence and all that." Another teacher-researcher spoke of the comments she had heard from male colleagues: "And now we're starting to get comments saying: 'well, basically, to get promoted you have to wear a skirt'" (UBx, TR).

In universities, the attachment to the neutrality of excellence would delegitimise positive actions aimed at women, but this point of view remains a minority one in the material collected.

### *An implicitly androcentric concept*

First and foremost, women (R or TR) pointed out the extent to which the great race implied by the imperative of scientific excellence was gender-biased and, to put it bluntly, unequal. Their experience echoes Sophie Lhenry's descriptions of the implicitly masculine norms of success in the academic profession (Lhenry,

2015). They are attentive to how gender and the social roles associated with being a woman shape their academic careers.

The availability implied by the vocational ethos of the research profession is much easier for men. Referring to their family responsibilities, the women who take part point to the necessary choices (between family life and work) that weigh more heavily on the women's side.

> Around me, girls often make the choice to sacrifice their personal lives by not having children, taking responsibility for it and saying: 'but if I want to get there, if I don't work day and night, if I don't take holidays – I won't be at the same level as my male colleagues'.
>
> *(UBx, TR)*

"I can confirm that. In my institute, in fact, the only groups of female leaders don't have children" (UBx, TR).

*Authority* is more easily exercised by men in academic organisations. One researcher recalls an anecdote that seems significant to her: when she was taking responsibility for a degree, her requests to her colleagues went unanswered until her colleague with joint responsibility (a male professor, she is a senior lecturer) relayed her request and obtained answers "within half an hour". The testimonies then multiply:

> yes, I'm going in the same direction *[pause]* it was for a professor post: so I was asked the question: was I aware of the responsibilities inherent in this type of post. I don't think my male colleagues were asked that question.
>
> *(UBx, TR)*

Finally, ambition seems more complicated for women. Some have the impression that their gradual specialisation in administrative tasks, which they explain by their gender, is preventing them from being academically productive.

> I have the impression that in the teaching teams I've seen, it's accepted as a fact that it's more women who take charge of a certain number of organisational, administrative and practical aspects of teaching, so afterward people say to me: 'yes, but that's because you do it better, because it's more straightforward, etc'. . . We want to do well for the students and so we end up doing these less noble tasks.
>
> *(UBx, TR)*

Others talk of being perceived as less ambitious, when in fact it's mainly domestic and family tasks that hold them back.

> So yes, it seems that we don't dare, that we're less . . . 'pushy' or we don't go as far as men. So that's why we don't dare to apply for these positions. But

> there's also the fact that if we take maternity leave, we don't necessarily have the same file.
>
> *(UBx, TR)*

When asked about the inequalities in scientific careers between men and women, most participants consider that the notion of scientific excellence is more suited to men as things stand.

## Equality as a prerequisite for excellence

In the dynamics of the discussion, elements of discourse emerge that shift the meaning of the relationship between equality and excellence, with equality becoming primary and a condition for achieving the latter. If excellence is only intended to produce a type of profile of excellent researchers, then it is devalued, and its validity depends on the actual achievement of equal opportunities. The dynamics of the discussion are particularly clear in the focus groups of leaders, teachers, and researchers at U.Porto. Starting with a definition of excellence based on a comparison with other universities, the group discussion clearly evolved towards an inclusive conception of excellence defined as the capacity to make equal opportunities effective. The definition of excellence is no longer external and comparative but internal, aiming for an "exemplary" work organisation that embodies universal values in a normative register. As one researcher put it, "There is no point in having good scientific indicators and a lot of material resources if, in practice, we have problems in terms of equality between men and women" (U.Porto, R.).

It is the ability to ensure a diversity of researcher profiles, particularly when recruiting, that is highlighted:

> We mustn't forget that neither men nor women all fit completely into the same mould, and I think that the more we can articulate a way of thinking that leaves room for this diversity of career paths . . . we gain by thinking about diversity for men and for women.
>
> *(UBx, TR)*

More generally, excellence is redefined in relation to equality from a moral or managerial perspective. "Let's assume that there are under-utilised groups who have qualities but who are not given the chance to exercise them. I don't think that would be excellence" (U.Porto, VD).

In this type of discourse, the excellence of universities must be embodied in their ability to develop equality of opportunity within them, to give credibility to the notion of scientific excellence, and to embody "universal values", a term that came up several times in the focus groups. The university ideal is that of an institution that ensures equality within itself so that it can disseminate universal values (such as merit) to the outside world. "It is also important to have an image that we transmit as a university to the outside world. How we treat our students, how we treat our colleagues, how we treat ourselves on these kinds of issues" (U.Porto, T.).

> Excellence is rather when we . . . well when our activities respect the values we have set ourselves, it's also a question of . . . hum . . . respect, ethics, respect for the whole community, which is in the university. For me, an excellent university is one that is not rated 20 out of 20 by criteria, but rather one where the community rates it 20 out of 20.
>
> *(UBx, AS)*

## Conclusion

Using focus groups to gather the opinions of staff at several European universities on scientific excellence and its relationship to equality, we have highlighted a number of points.

Over and above the differences in context that shape the discourse, the criticism of scientific excellence is severe and can be found in each of the universities, particularly among teachers and/or researchers. Scientific excellence appears to be an ideal that has gone astray and is perceived as putting pressure on researchers, at the risk of distancing them from their professional values, which requires time.

A very small number of people (male researchers) do not consider it relevant to take inequalities into account and adhere to the neutrality of excellence criteria, which they believe guarantee equality. The vast majority of speakers develop a critique of excellence, highlighting the way in which men are more likely to achieve these criteria given the gender regime in society but also in academic organisations, and deplore the way in which these criteria lead to a single model of researcher. In view of the results obtained by Nentwich et al., it is not so much a question of the deficiencies of women in relation to a model held to be universal, but rather of highlighting the "masculine" character of the model of the "excellent" researcher. This shift may be indicative of the circulation and appropriation of feminist work on gender in academic circles.

Finally, in the course of the discussion, a reversal of the relationship between excellence and equality can be observed among certain groups (particularly in governance): equality becomes primary and justifies, in a normative sense, institutional change to restore the ideal. If excellence corresponds only to a small, socially situated group, then the notion of merit tends to be lost, and the effective diversification of researchers considered excellent restores the notion's meaning. This articulation, conscious of the social handicaps assumed by certain groups with regard to excellence, paves the way for significant institutional change by making possible a policy of equal opportunity or equity for scientific recruitment or promotion. The exemplary role the university wishes to play in society would then require the effective diversity of its staff, particularly researchers, to embody the "universality" that universities claim for themselves.

The results obtained must be seen in the light of the identity of the focus group participants: people who volunteered to think about equality issues at their university and try to influence the course of events. They are not necessarily representative of all employees. As a result, the comparative dimension emphasised the similarities between the discourses heard at the different universities studied, and failed to take sufficient account of the differences between professions and countries. Why, for example, do administrative staff at UBx and U.Porto take up the issue of excellence, while administrative staff at AUTh and UL do not? A few contextual elements presented in the introduction shed some light, but the question remains to be explored. The views of administrative staff at UBx and U.Porto on excellence clearly show that in these universities, the imperative shapes administrative work. This notion does have a collective dimension for universities (without even mentioning the impact of rankings), even if, as in this chapter, it primarily concerns the individual dimension of research.

## References

Augoustinos, Martha. (2005). New racism, meritocracy and individualism: Constraining affirmative action in education. *Discourse & Society, 16*(3), 315–340. https://doi.org/10.1177/0957926505051168

Fassa, Farinaz. (2015). Excellence and gender: Playing the game of scientific excellence or being played by the game? The Swiss example. *Equality, Diversity and Inclusion: An International Journal, 34*(1), 37–54.

Ferree, Myra Marx, & Zippel, Kathrin. (2015). Gender equality in the age of academic capitalism: Cassandra and Pollyanna interpret university restructuring. *Social Politics: International Studies in Gender, State & Society, 22*(4), 561–584. https://doi.org/10.1093/sp/jxv039

Lhenry, Sophie. (2015). Les normes implicites de réussite. In Rebecca Rogers & Pascale Molinier (Eds.), *Les femmes dans le monde académique*, 95–106. Presses universitaires de Rennes.

Musselin, Christine. (2017). *La grande course des universités*. Presses de Sciences Po.

Nentwich, Julia, Heydenreich, Anna-Kathrin, & Offenberger, Ursula. (2019). When excellence meets equal opportunities: Discourse analytical findings. *Sociologies*. https://doi.org/10.4000/sociologies.11932

# 4

# RESET'S COMMITMENT TO INCLUSIVE SCIENTIFIC EXCELLENCE: MOBILISING GOVERNANCE AND INCLUDING NEW INDICATORS

*Ninon Junca*

## Introduction

One of the aims of the RESET project is embedded in its title: to redesign scientific excellence from a collaborative perspective, making it more inclusive and adaptable to contemporary challenges.

To achieve this, top management members of the seven RESET partner institutions co-designed and endorsed a Joint Statement in 2022, demonstrating their commitment to equality, diversity, and excellence in research. This unprecedented bold statement in favour of transforming the governance, culture, and missions of universities towards equality highlights the false neutrality of scientific excellence and sets out the conditions for making it fairer. The primary purpose of this document is to leverage insights from critical feminist studies to foster institutional change in the conception of scientific excellence within universities and the European Research Area (ERA). This chapter aims to present the content of the statement, its operationalisation through a series of indicators, and the mobilisation method on which it is based. Its objective is to provide guidance to stakeholders involved in scientific excellence policies within institutions, national funding agencies, and the ERA. It is based on the genesis of a joint commitment by seven universities and a case study (the University of Bordeaux) for concrete operationalisation – between the ambitions of such a political document and what is feasible in terms of metrics follow-up and institutional change of practices.

### *The RESET joint statement, a tool for institutional change*

The statement outlines four main areas of action aimed at promoting sustainable institutional and cultural change towards greater equality and diversity, while

DOI: 10.4324/9781003564348-7

advancing the production of excellent research and innovation outputs. This document illustrates RESET members' joint conception of a more inclusive scientific excellence, namely by its social and environmental dimensions, through engagements anchored in four thematic areas:

The RESET joint statement acts upon three levels:

- The constitutional and institutional level – framing RESET universities' culture towards equality and diversity;
- The operational level – encouraging the effective implementation of concrete actions to change practices towards more inclusion;
- The decision-making level – enabling the development of institutional policies and core documents to underpin these principles and ensure their endorsement.

The purpose of this document is to develop and adopt inclusive institutional practices and frameworks of recruitment, career progression, work–life balance, culture, and excellent research schemes. Each of its aspects is accompanied by indicators aimed at reflecting about a more inclusive excellence assessment, and ensuring the effective integration of the statement's principles at the institutional level. This document offers a paradigm shift in the understanding of scientific excellence.

The RESET Joint Statement is the result of a collective effort conducted in four stages: qualitative data collection and literature review, collaborative reflection and capacity-building sessions to define a shared vision of scientific excellence, a decision-making session, and the selection of common indicators. The third stage involves participation of university stakeholders including top management

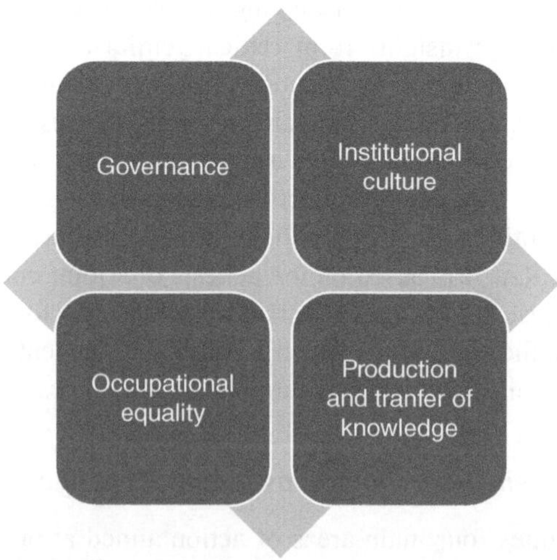


**FIGURE 4.1** Four thematic areas

members: Presidents, Rectors, and Vice-Rectors of seven partner universities of the RESET project.

### *Main findings of qualitative data collection and literature review that resulted in the statement*

As illustrated in previous chapters, the concept of scientific excellence is contested, with its definition and relevance being questioned within academia. Scientific excellence is often assessed via some recurring criteria that are supposed to be neutral, but presume gender-blindness: productivity, peer-reviewing, number of quotations, and international recognition of researchers (Van den Brick & Benschop, 2012).

Moreover, the definition of excellence is not universally agreed upon, even at the highest levels of policymaking, and should be adapted according to the local context and stakeholders (O'Connor & O'Hagan, 2016; Nielsen, 2015; Ferretti et al., 2018; O'Connor & Barnard, 2021). However, excellence can be measured with the help of indicators, whose construction typically involves activities of quantification, as research excellence has no physical substance in itself (Ferretti et al., 2018).

At RESET institutions, the perception of scientific excellence was explored through focus group sessions with members of top management, researchers, teachers, and staff (see Chapter 2) and via a survey disseminated to all personnel of the RESET universities. It has also been a topic of one of the meetings of Gender Equality Boards at UBx, RUB, and U.Porto. In addition to that, the UBx RESET team conducted a set of individual interviews with different institutional stakeholders related to the excellence funding schemes.

Institutional stakeholders interviewed at UBx emphasised the importance of the collective dimension of excellence. It is seen as crucial to value the efforts of teams to achieve excellence, taking pedagogical duties into consideration and encouraging diversity and equality in research teams. Multidisciplinarity and sharing of experiences and knowledge among people with different social and cultural backgrounds empower effective and impact-driven innovation.

The discussions run with Gender Equality Boards went in the same direction. GEB members of U.Porto stressed that it is important to take into account specificities of scientific fields. Some criteria seem to be disadvantageous for women due to the difficulty of finding a balance between professional life and family duties. These elements were also supported at the sessions of GEBs in RUB and UBx. At the University of Bordeaux, members of the GEB also emphasised the importance and will to focus not only on scientific excellence but also on academic excellence: the working and studying environment, quality of teaching and research content, and promotion of equality, diversity, and societal responsibility. They should be central in the evaluation of local excellence. However, according to them, it is neither possible nor desirable to renounce the position of the institution within international rankings and in international excellence schemes. During their meeting, GEB members of RUB developed some ideas on how to improve excellence towards equality and diversity in recruitment and advancement procedures, such as enabling more transparency and communication for candidates.

To sum up, the concept of scientific excellence itself cannot be undermined and will always be relevant for decision-makers of higher education institutions, as it drives change and innovation. It is, thus, the responsibility of higher education institutions to adjust their cursor and question their criteria, in order to adapt them to societal challenges.

This chapter presents guidelines towards scientific excellence and a discussion on indicators to measure it – to which the RESET university top management members have committed themselves. It also illustrates the methodology used to draw up this document and the way its principles have been implemented at local institutions.

## Content of the RESET joint statement

Through this joint statement, the RESET team members aim to share their vision of scientific excellence and contribute to the progressive change of its assessment: to make its criteria more inclusive and diverse, in accordance with social challenges, as well as with the diversity of people who produce it. This document highlights six main criteria that must be taken into consideration when thinking about inclusive scientific excellence: recognition, multidisciplinarity, open access, intersectionality, responsible research and innovation, and societal impact.

Signatories of the statement commit to act upon several thematic areas:

- **Governance**: by strengthening their institutional engagement to equality and diversity principles through the review of core texts and official guidelines; enhancing co-design and collaboration in the creation, development, and monitoring of ideas and actions in favour of equality and diversity; and by fostering diversity and equal representation in decision-making bodies and processes.
- **Institutional culture:** by promoting sustainable cultural change through the establishment of a gender and diversity-friendly environment and communication; integrating gender mainstreaming in their policies; and combating all forms of discrimination or gender-based violence.
- **Occupational equality**: by refining recruitment, retention, and decision-making processes, in line with HRS4R and European standards; reflecting on the impact of parenthood on careers and solutions to enable a work–life balance; and valuing the contribution of administrative and academic communities within European responsible research and innovation.
- **Production and transfer of knowledge**: through the development of training programmes and pedagogical content that raise awareness to equality and diversity issues; the implementation of the Gender Impact Assessment (GIA) tool

of RESET; and the development of societal and environmental dimensions of research.

## *Political dimension of the RESET joint statement*

It was crucial that top management of RESET institutions participate in the reflection on scientific excellence and that they sign the statement to ensure its institutionalisation. Indeed, a diversity policy only comes under the scope of change management when it is part of a strategic, systemic, cross-functional, and sustainable approach (Bruna, 2013). To achieve this, it needs to be endorsed at an institutional level by decision-makers who embody the university's culture. By participating in the RESET project, the top management of the seven institutions had already demonstrated their commitment to equality and diversity. The signing of the statement represents a significant step forward in the realisation of this commitment.

At a crossroads between research and human resources, top management members know well their institutions and the way to enshrine the principles of equality and diversity over the long term in our definition of scientific excellence. To ensure institutional support, the statement had to be signed at the highest political level, with variations depending on the local organisation: president, director, rector, and/or vice-rector.

After combining the results of qualitative analysis (focus groups, surveys, interviews, and literature review), the team in charge of the elaboration of the statement (UBx) organised a meeting with representatives of top management from all RESET institutions in March 2022, in order to agree on the document. It was an occasion for Rectors and Vice-Rectors to meet and to exchange ideas on this challenging topic, sharing similar thoughts. Besides the fact of agreeing on the content of the document to sign, this meeting had an important symbolic value for RESET institutions' top management members' engagement towards equality and diversity. The document was then signed by the representatives.

It was translated into French and Polish. The statement principles have been integrated into the elaboration of Gender Equality Plans and the development of actions at a local level. It is publicly available on the local websites, the RESET website, and Zenodo[1] to ensure its dissemination at the ERA level.

## *Elaboration of indicators to ensure monitoring, endorsement, and sustainability of the RESET joint statement*

After the signature and dissemination of the joint statement, RESET members reflected on the best way to ensure the endorsement of its principles by all institutions, giving it a more coercive dimension. It was agreed to select a shortlist of realistic and relevant indicators that all RESET institutions would engage in following even after the end of the project's lifespan. The University of Bordeaux team was tasked with this effort and developed a general list of 83 indicators, taking into account the four thematic areas and the literature review. Afterwards, it was accompanied by the Economic Analysis and Statistical Studies Department of the University of Bordeaux to refine metrics and select a shortlist.

The following questions were asked:

- Is the indicator already followed up at the university, and by whom?
- If it is not yet followed, is it really useful to follow it?
- If yes, is it feasible to collect this data, and who will be responsible for it?

This reflection was completed by an analysis of other lists of indicators that the University of Bordeaux has to follow, such as the ones for the label on Sustainable development (DD&RS), or national calls for projects. The UBx team could also count on the support of various institutional services to reflect on the relevance and feasibility of metrics follow-up.

In this chapter, we present the example of the indicators selected by the University of Bordeaux, also illustrating the limits of monitoring with metrics. Even in an excellent university that is strongly engaged towards equality and diversity, it is sometimes challenging to collect data and to ensure the follow-up of the implementation of an engaging collective document such as the RESET joint statement. Each partner of the RESET project has taken the UBx work on indicators as a starting point to implement it at a local level. Thus, part 2 is a case study designed to be useful for institutional decision-makers who would like to redesign their definition and assessment of scientific excellence. However, it is important to highlight the fact that implementation of such an engagement depends on the local context and institutional specificities, and thus, adjustments may be necessary.

## The challenge of selecting indicators for the operationalisation of a policy document: example of the University of Bordeaux

### *Selection of indicators*

The primary purpose of follow-up indicators is to monitor the implementation of policy effectively. However, to maintain their relevance, indicators must be selected carefully and strategically. Their number should be limited to ensure the feasibility of follow-up and to distribute the workload responsibly among human resources.

To achieve this, the University of Bordeaux RESET team collaborated with the Economic Analysis and Statistical Studies Department to develop a shortlist of indicators based on the statement's four thematic areas. The following section will illustrate the challenges involved in monitoring and operationalising a collective and binding document, and how it was adapted to the local context.

### *Indicators chosen for the monitoring of thematic area 1: governance*

The institutional support from the governance is an essential condition for legitimating and highlighting the significance of inclusion, societal impact, and collective efforts within excellence (Joint statement of top management on their engagement for equality, diversity, 2022).

RESET institutions believe that it is important to ensure a gender balance in the composition of boards and committees whose members are in charge of decision-making and strategic choices, such as recruitment. It is also relevant to monitor training and communication activities targeting issues of diversity and equality in decision-making. While equality has nearly been reached at the level of PhD graduates, only 24% of top positions in the higher education sector in the European Union are occupied by women (European Commission, 2020). Thus, the main criteria for the governance area is to observe equality in decision-making bodies, which can be facilitated through a dedicated national and local policy (statutes, codes, and charters) on gender balance in the administrative and academic councils, commissions, peer review and selection committees, and other structures of this type. In the case of the UBx, the national regulations are a lever to ensure parity in recruitment boards (a minimum of 40% of the under-represented sex in academic recruitment boards). This parity objective can be completed by the existence and monitoring of implementation of training activities on gender equality, diversity, biases, and stereotypes for members of selection committees. To illustrate this evidences, selected metrics by the UBx are:

- Share of women/men in promotion or recruitment boards and in decision-making bodies by scientific field.
- Share of women/men in academic and administrative decision-making positions (e.g., boards, councils, committees, and juries).
- Gender of the head of the institution and their political teams (Vice-Rectors/Presidents).
- Sex-disaggregated data on training sessions for selection committees.

In this thematic area, there was no difficulty in selecting these indicators that are already followed up by the human resources services at the University of Bordeaux, and framed by the national law.

### *Indicators chosen for the monitoring of thematic area 2: institutional culture*

The institutionalisation of diversity policy, which marks a bureaucratic shift, guarantees its official recognition and sustainability (Bruna, 2013). It also encourages structural change in accordance with sustainable models of development and innovation (Joint statement of top management on their engagement for equality, diversity, 2022). Thus, it is important to ensure that principles of diversity and equality are visible and promoted in the institutional culture, through practices, rules, communication frameworks, and environment.

This thematic area has a large scope and is composed of criteria in line with the RESET vision on the sustainable and impactful institutional and cultural change, namely through recruitment, career advancement, gender-inclusive communication, training, and recognition of the collective dimension of research.

The main selected criteria for the "Institutional culture" are the following:

**Transparency of institutional systems**

Transparency is a key element of institutional culture since it enables trust in the objectivity of the institutional decision-making processes (European Commission, 2019). The evidence selected for this area is the availability and publication of sex-disaggregated data on teachers, researchers, staff, and students (by age group, scientific field, and (dis)ability), and on cases of gender-based violence and discrimination. Additionally, to the data on the reporting unit (if it exists), implementation of a qualitative survey on discrimination and violence at the university can be a good way to collect the university community's point of view on the topic and raise awareness of this issue.

**Policy for equality and diversity**

This criterion focuses on the existence of institutional policies, documents, and facilities that favour the establishment of an equal, safe, and inclusive environment. According to the results of the Emerald Global Inclusivity report[2] (2022), 90% of academics agreed that inclusivity promotes diversity of thought, 88% consider that it creates a more open learning culture, and 87% agree that it has a positive effect on creative thinking.

Thus, HEIs should dedicate resources and put effort into developing and/or improving existing policies and guidelines contributing to the establishment of an equal and diversity-friendly environment. The quality of content and efficiency of these policies, as well as their actual implementation, must also be monitored.

The selected evidence criteria chosen for this are the following:

- Existence of an institutional structure/unit to promote gender equality, diversity, inclusion, and a system for monitoring its actions.
- Existence of a procedure for the dissemination of strategic documents, including values of equality and diversity, to all new employees and students.
- Existence of a page dedicated to equality/diversity on the institutional website.
- Existence of a policy for tackling gender-based violence and discrimination (procedure, protocol, unit, and portal).

- Elaboration and dissemination of reports on the number of cases of gender-based violence and discrimination (number of referrals and treated cases of reported gender-based violence and discrimination. Sex-disaggregated data on reported cases).
- Existence of a policy on gender-inclusive and/or diversity-oriented communication and a system for monitoring its implementation.
- Existence of training activities and tools on GE, diversity, and inclusion (sex-disaggregated data on personnel/students who followed training, by field/unit).

### Equal recognition in the attribution of prizes, awards, and funding

Excellence in research is often recognised through the attribution of specific prizes. However, the criteria for evaluation of candidates should consider potential differences in the career paths and research fields of the selected candidates or applicants. Moreover, institutions can install gender-sensitive policies that will positively contribute not only to their image but also to the establishment of a more inclusive and safer environment. At the University of Bordeaux, there is no centralisation of sex-disaggregated data on the attribution of prizes, awards, or funding. It depends on the research unit or institution in charge of it. However, with the reflection on the RESET Joint Statement, it was possible to ask the dedicated service to follow up sex-disaggregated data on laureates of Doctor Honoris Causa. Additional transparency should be ensured with the centralisation of data for other prizes.

### Environmental responsibility

At the University of Bordeaux, there is a Transition Charter that some laboratories signed. It is an excellent example of how societal and environmental dimensions should be jointly considered in the evaluation of the excellent academic environment. The signature of this document has two main objectives: contemplation of the working conditions and excellent scientific research schemes and setting up of objectives that will lead to an action plan. According to the Charter's strategy, the heads of research departments must encourage research units and laboratories to endorse it. The dissemination of the Charter should be accompanied by a guide that facilitates monitoring of indicators and presents best practices.

Universities and their communities are not isolated structures; humans are the main driving force of the Anthropocene: they exist and impact ecosystems. Some academics even suggest applying the Index of Sustainable Development to the evaluation of the excellence of universities (Hamant & Jensen, 2022). In order to implement principles of equality and diversity and to answer contemporary challenges, it is fundamental to adopt a wider strategy of social and environmental transitions. Moreover, gender inequalities are directly reflected in the access and usage of environmental resources (Forest & Foreste, 2022), and the environmental issue is the current setting of reflections on gender inequalities through the question of care (Laugier, 2015). Gender equality is inscribed as Sustainable Goal 5 of the United Nations.

At the University of Bordeaux, there are various strategy documents that combine issues of environmental and societal responsibility. Their existence is positive evidence of the endorsement of the statement principles and can be used as an indicator by other institutions. The follow-up of the environmental footprint of personnel, research units, and campuses is the main indicator selected for this area.

### *Indicators chosen for the monitoring of thematic area 3: occupational equality*

Occupational equality – in academic and non-academic career paths – should be reflected in the institutional policy for recruitment and advancement, strategy for equal payment and internationalisation, support of teaching and research-related activities, and measures towards balance between work/studies and personal life. The process of providing employees and students with all the elements necessary for their fulfilment, dedication, and growth should be carried out considering the balance between professional and personal life, since excellent results require excellent environments (Joint statement of top management on their engagement for equality, diversity, 2022).

The university can support its equality and diversity strategy by ensuring transparency on the following elements, and their recognition in career advancement criteria:

- Heterogeneity of social characteristics of personnel (by collecting intersectional data such as age, type of contract, origin, (dis)ability, and scientific field).
- Work–life balance: sex-disaggregated data on persons applying for/taking parental leave; duration of the leave; and existing facilities (e.g., home office, maternity leave, and paternity leave).
- Sex-disaggregated data on use of work–life balance and care-related facilities.
- Pedagogic innovation and dedication in the evaluation and selection of candidates (sex-disaggregated data on leaves dedicated to the participation in and elaboration of pedagogical projects).
- Recognition of contributions made by the research support offices (monitoring of calls for projects, writing of proposals, administrative and financial support, and project management) through dedicated media campaigns, eligibility for funding, or inclusion of their names to the final outputs.
- Existence of tools and processes that facilitate international mobility (guides, doctoral networks, mentoring programmes, and availability of information in several languages).

The advantage for the RESET UBx team is that in France, there are numerous legal frameworks about the annual production of data on public service workers. These indicators (sex-disaggregated data on type of contract, promotions, recruitment, work–life balance facilities, etc.) are thus already followed up and presented in a public document that is voted by the Administrative Board every year.

### *Indicators chosen for the monitoring of thematic area 4: production and transfer of knowledge*

The development of training programmes, pedagogical content, and methodologies that bring awareness to equality and diversity issues is essential in order to ensure viable and long-term effects. Gender mainstreaming in research and teaching activities not only enables effective awareness-raising on these issues but also contributes to the innovation and potential for excellence of the institution. Gender mainstreaming in pedagogical and research activities is further promoted through dedicated actions in local Gender Equality Plans (AUTh, UBx, UL, and U.Porto).

The collection of sex-disaggregated data on the funding of projects improves data-driven and more objective decision-making and distribution of resources. The integration of knowledge on gender inequalities into the teaching and training activities contributes to the sustainability of the gender equality and diversity policy.

According to the European Innovation Council Pilot Advisory Board (2021), it is essential to promote gender equality in research and entrepreneurship and to achieve balanced participation in the European Innovation Council supported activities, in order to "win the race to lead the next wave of strategic technologies".

The indicators identified by the UBx team to monitor the integration of gender and diversity dimensions in research and teaching activities are the following:

- Existence of a GIA policy document (protocol and checklist of RESET) and/or training for applicants about the integration of the sex and gender dimensions in research, and system for monitoring its dissemination.
- Existence of a system for the monitoring of courses and tools that help to integrate gender and diversity dimensions (number of hours of dedicated institutional courses; and number of Bachelor, Master, and PhD programmes and training classes that explicitly integrate gender equality and diversity in their titles and/or contents).

At the University of Bordeaux, data on courses and their content related to societal and environmental transitions is still not available but various services are working on this to bring more transparency to these strategic issues.

Last but not least, it is important to monitor the social impact of knowledge production and dissemination. As universities and research structures receive public funding, their outputs and infrastructure should contribute to the dissemination of socially valuable knowledge, favour innovation and the maintenance of a dialogue between science, research, and citizens (participative science) as well as support Open Science practices and principles.

The scientific and academic services (e.g., scientific mediation and communication, laboratory services, and training activities delivered for external audiences) provided by the universities should be taken into account in their evaluation for the distribution of public funding and awarding labels that imply excellence (Barrere, 2020).

Thus, the following indicators have been selected:

- Existence of services and mechanisms that enable participative science (number of EU, national, and local projects that enable participative science).
- Existence of institutional policy on Open Science and mechanisms for its monitoring.

These are the indicators identified by the University of Bordeaux team as relevant and possible to collect for the monitoring of inclusive scientific excellence. Discussions are still ongoing in order to determine the frequency with which they will be collected and the way they will be highlighted (publication, analysis, etc.).

The presentation of selected criteria and indicators targets the improvement of monitoring and the evaluation of scientific excellence. The process of reflection must be ongoing in order to provide more inclusive indicators and criteria for the evaluation of both individual and institutional levels of scientific excellence. The RESET project members thus ended with a further phase of debate, discussion, and co-design, leading to an updated version of the statement.

### *The limits of indicators*

The selection and adoption of indicators have been much more challenging than the elaboration of the document in itself. Indeed, the diversity of local contexts and monitoring tools problematised the selection of common indicators. At a local scale, we can observe a general fatigue in relation to indicators, at a time when institutions are overwhelmed with demands of metrics on various topics linked with productivity and sustainable development. National and international labels such as HSR4R are also very demanding in terms of monitoring and production of metrics. Some authors even speak of a race for indicators (Déjean, 2021).

The difficulty of finding common indicators is also evident in the absence of an international reference method for constructing indicator systems and methodologies (Descamps & Vicard, 2010). It is challenging to plan how to centralise data and to determine who will be in charge of it.

We can observe goodwill from universities' members to follow-up indicators, but the implementation proves challenging because of three main elements: there is a lack of human resources support in laboratories that are in charge of collecting a large proportion of indicators, where there are very few positions for this task, which results in a heavy workload. Furthermore, follow-up of indicators can be seen as a source of constraint and their usefulness can be questioned when stakeholders do not necessarily have control over the elements that can make these indicators evolve in a more positive way. There is a need for support that is not always possible. Finally, the policies promoted by the central level of universities may encounter relative inertia within the laboratories, which are central to the process, and sometimes resistant to the production of new indicators. In the very specific case of UBx, this inertia is also due to the fact that some laboratories depend on multiple administrative supervisions (the UBx shares their management with other national or regional research centres). This implies a variety of operating means, methodologies, and indicators.

In the case study of the University of Bordeaux, the team had to put aside some indicators previously identified as relevant, since it was not possible to collect them.

The first challenge is related to the sex-disaggregated data on research funding. Indeed, existing tools do not enable the production of such indicators on research funding and publications. Practices and methodologies may vary from unit to unit, institution to institution, or country to country. The issue then becomes how to ensure more transparency while safeguarding workers' well-being, knowing that the production of additional indicators may be time-consuming. This indicator would be relevant to understand the potential gap between men and women on scientific productivity, which would explain the differences in access to excellence recognition.

Moreover, there is no system for an open and transparent advertisement of positions in research (expected skills and experience, criteria for selection, and description of accessibility). At the University of Bordeaux, it is not yet possible to follow up these criteria in a very effective way, so they were removed from the shortlist. Even if the institutional policies changed, in practice, transparency is not always observable.

At the University of Bordeaux, we do not have indicators on nominees nor gender distribution among keynote speakers or members of the jury at conferences, since this data is not centralised and depends on the organisers. There is a lack of transparency in the processes of awards since it is very research unit-dependent: the institutional commitment to promote parity at all levels (highlighted by the actions of the Gender Equality Plan) is not necessarily known or implemented by research units or laboratories. The only data we could engage in following is that on Doctor Honoris Causa, since it is very limited in terms of number.

## Conclusion

The RESET Joint Statement represents an unprecedented and exemplary commitment from the top management of seven leading European universities. This initiative marks a bold and deliberate effort to reshape long-standing practices in research assessment and scientific excellence, with the shared goal of fostering a more inclusive, equitable, and diverse academic environment. The ability of these institutions to agree on and commit to common principles and actions underscores the significant political and symbolic value of this initiative.

Operationalising the principles of the RESET Joint Statement through the selection of specific indicators is a crucial step in demonstrating the genuine engagement of these institutions. These indicators not only provide a tangible means to monitor progress but also serve as a framework for accountability, ensuring that the commitment to inclusivity is translated into concrete actions and measurable outcomes.

However, the process of implementing and monitoring these indicators has revealed significant challenges, even within institutions that are deeply committed to transparency and evolution. Key obstacles include the scarcity of human resources, the heavy workload associated with data collection and analysis, and the centralised nature of decision-making processes. These factors can hinder the ability of university stakeholders to fully embrace the statement's principles and consistently track a large number of indicators.

To address these challenges, it is essential to limit the number of indicators to those that are most relevant and feasible, ensuring that the follow-up process remains manageable and effective. At the same time, there is a pressing need to develop and refine tools for tracking metrics, particularly in areas such as research funding, where sex-disaggregated data transparency is critical.

The experience of the University of Bordeaux, which serves as a model for the other RESET partners, highlights both the opportunities and the complexities involved in making scientific excellence more inclusive. The process of changing deeply ingrained practices requires not only institutional commitment but also flexibility and innovation in the face of challenges. As the RESET initiative moves forward after the project lifespan, it will be crucial to continue refining these strategies, fostering collaboration, and sharing best practices across institutions to achieve the shared vision of a more inclusive and equitable academic landscape.

## Notes

1 https://zenodo.org/records/14646461.
2 https://www.emeraldgrouppublishing.com/global-inclusivity-report-2022.

## References

Barrere, Rodolfo. (2020). Indicators for the assessment of excellence in developing countries. In *Transforming Research Excellence: New Ideas from the Global South*. African Minds. https://doi.org/10.5281/zenodo.3603891

Bruna, Maria Giuseppina. (2013). Le portage d'une politique diversité comme processus de changement organisationnel: phénoménologie et jeux d'acteurs. *Question(s) de Management*, *2*(3), 25–35. https://doi.org/10.3917/qdm.132.0025

Déjean, Frédérique. (2021). Responsabilité sociale de l'entreprise et performance intégrée: la course aux indicateurs. *L'état du Management*, 37–48. ⟨hal-03481704⟩

Descamps, Mélanie, & Vicard, Fanny. (2010, octobre). Mesurer le développement durable: jeux d'indicateurs et enjeux locaux. *Revue d'Économie Régionale & Urbaine*, (4), 749–771. https://doi.org/10.3917/reru.104.0749

European Commission: Directorate-General for Research and Innovation. (2019). *ERA progress report 2018: The European research area: Advancing together the Europe of research and innovation*. Publications Office. https://data.europa.eu/doi/10.2777/118067

European Commission: Directorate-General for Research and Innovation. (2020). *A new ERA for research and innovation: Staff working document*. Publications Office. https://data.europa.eu/doi/10.2777/605834

European Innovation Council pilot Advisory Board. (2021). *Statement on gender and diversity in EIC*. https://eic.ec.europa.eu/document/download/8d2b88ac-2516-4f03-bc10-f3e5641ecdf1_en?filename=EIC_Gender_Diversity%20statement_June%202021.pdf

Ferretti, Federico, Guimarães Pereira, Ângela, Vértesy, Dániel, & Hardeman, Sjoerd. (2018). Research excellence indicators: Time to reimagine the 'making of'? *Science and Public Policy*, *45*(5), 731–741. https://doi.org/10.1093/scipol/scy007

Forest, Maxime, & Foreste, Christlord. (2022). Quels cadres interprétatifs autour des enjeux Genre et Climat? Enseignements d'une analyse bibliométrique. In H. Périvier & S. Rabier (Eds.), *Quels cadres interprétatifs autour des enjeux Genre et Climat: Enseignements d'une analyse bibliométrique*, 1–104. Paris: Agence française de développement. https://www.cairn.info/papiers-de-recherche-1000000148989-page-1.htm

Hamant, Olivier, & Jensen, Pablo. (2022, September 8). Universités: "Remplaçons le classement de Shanghaï par un palmarès plus durable". *Le Monde*. https://www.lemonde.fr/sciences/article/2022/09/08/universites-remplacons-le-classement-de-shanghai-par-un-palmares-plus-durable_6140652_1650684.html

Laugier, Sandra. (2015). Care, environnement et éthique globale. *Cahiers du Genre*, (59), 127–152. https://doi.org/10.3917/cdge.059.0127

Nielsen, Mathias. (2015). Gender inequality and research performance: Moving beyond individual-meritocratic explanations of academic advancement. *Studies in Higher Education*, *41*(11), 2044–2060. https://doi.org/10.1080/03075079.2015.1007945

O'Connor, Pat, & Barnard, Sarah. (2021). Problematising excellence as a legitimating discourse. In *The Routledge Handbook of Gender and Excellence*, 37–50. https://doi.org/10.1007/978-3-030-69687-0_3

O'Connor, Pat, & O'Hagan, Clare. (2016). Excellence in university academic staff evaluation: A problematic reality? *Studies in Higher Education*, *41*(11), 1943–1957. https://doi.org/10.1080/03075079.2014.1000292

Van den Brick, Marieke, & Benschop, Yvonne. (2012). Gender practices in the construction of academic excellence: Sheep with five legs. *Organisation*, *19*(4), 507–524. https://doi.org/10.1177/1350508411414293

# PART II

# Mobilisation and resistance

# INTRODUCTION TO PART II

## Mobilisation for inclusive excellence, confronting resistance to rethink equality and excellence together

This section advances the discussion on scientific excellence and equality by reflecting on the development of transformative instruments for higher education institutions (HEIs) and showcasing practical tools to mobilise communities, confront resistance, and assess change. The first chapter by Livari et al. explores the potential of co-design as a mechanism for promoting gender equality, drawing on three empirical applications within the RESET project. The use of these approaches to critique the status quo, explore alternative futures, and provoke political and social debate is discussed. In addition, four principles of co-design in gender equality work are defined: creating safe spaces for addressing power dynamics, encouraging criticism and exploration of alternative futures, fostering organic and inclusive participation, and leveraging participatory infrastructure for a sustained social change. These principles align co-design with ethical and political aims of gender equality work, advocating for transformative and long-lasting impacts. In the following chapter, Matias et al. examine the role of GEPs in addressing gender inequalities, describing the design and implementation process of two iterations of GEPs within RESET. A critical perspective on efficiency-based justifications and conservative frameworks is presented to highlight institutional resistance and the challenges of adopting diversity and intersectional approaches. The authors emphasise the need for innovative and adaptive strategies to broaden the equality agenda and meet the evolving needs of academic institutions.

The subsequent chapters focus on practical tools that foster institutional and cultural change in universities. Michali et al. explore the obstacles academic and research institutions face in managing diversity, equality, and inclusivity (DEI)

DOI: 10.4324/9781003564348-9

data, focusing on issues of accuracy, reliability, ethics, and institutional barriers. The authors also provide practical recommendations for data collection, serving as a valuable resource for practitioners, policymakers, and scholars committed to advancing inclusion in academia. The final chapter of the section by Viktoria Niebel examines the implementation of gender-inclusive and diversity-sensitive communication practices in academic institutions, highlighting the critical roles of verbal and visual language in shaping representation and inclusivity. Drawing on insights from the RESET Toolbox, this chapter outlines strategies to counter resistance, promote safety, and address intersectional barriers. It advocates for transformative communication practices that foster genuine inclusion and a sense of belonging within higher education.

Overall, the second part of this volume demonstrates how universities can be mobilised through practical strategies and tools to achieve a more inclusive excellence, emphasising diversity and equality as central components of institutional transformation.

# 5

# CO-CULTIVATING CO-DESIGN FOR GENDER EQUALITY IN THE CONTEXT OF HIGHER EDUCATION INSTITUTIONS

*Netta Iivari, Eva Durall Gazulla, and Ruut Tikkanen*

## Introduction

We have been engaged in co-cultivating co-design to address gender equality in Higher Education Institutions (HEIs) in the European Union (EU) funded RESET project. The promotion of gender equality is currently on the agenda in the EU. The European Commission (EC) shows its commitment to the promotion of gender equality in research and innovation in various ways: for instance, through the EC's Gender Equality Strategy and by making gender equality a priority in the Horizon Europe research and innovation funding programme, in which the Gender Equality Plan is used as an eligibility criterion for applicants, including HEIs. There is a requirement to integrate the gender dimension into research and innovation content which is to be evaluated under the excellence criterion, and there is an explicit objective to increase the gender balance in Horizon Europe-related boards, evaluation committees, and research teams (EC, n.d.). However, there are still many challenges in the promotion of gender equality in research and innovation, including in HEIs: the gender gap still prevails regarding women's and men's participation in different research fields and hierarchical levels in HEIs, and there are many political, cultural, and organisational obstacles that inhibit efforts to move in a more equal direction (e.g., Iivari et al., 2022; EC, 2021).

Co-design has been proposed as a solution to tackle some of the challenges. Co-design, referring to "collective creativity as it is applied across the whole span of a design process" (Sanders & Stappers, 2008, p. 6), invites stakeholders to envision and specify design solutions in a collaborative manner. For promoting gender equality in HEIs, it is assumed that it is valuable to invite stakeholders (staff and students) to co-design new tools, practices, and policies advocating gender equality. While we perceive a great deal of value in co-design in this context, we

DOI: 10.4324/9781003564348-10

think there are critical aspects neglected in mainstream co-design that, due to its historical origins, could be effectively integrated into it and that would prove particularly valuable in the context of gender equality work. With this study, we advocate greater acknowledgement of the participatory and critical roots of co-design, derived from the Scandinavian tradition of Participatory Design (e.g., Greenbaum & Kyng, 1991) as well as from various traditions and forms of critical and speculative design (e.g., Bardzell & Bardzell, 2013; Dunne & Raby, 2013; Iivari et al., 2022) and from futures-making and anticipation traditions (Amsler & Facer, 2017; Coleman & Jungnickel, 2023; Slaughter, 2018; Whyte et al., 2022).

This chapter discusses the ways in which we have co-cultivated co-design by integrating more participatory and critical approaches into co-design for the promotion of gender equality in HEIs. We describe the theoretical background of the study, our methodological choices, and empirical examples from RESET, in which we have integrated more critical and participatory approaches into co-design. Based on our empirical experiences, we propose principles for co-design for gender equality – ones that offer a novel approach not yet elaborated in the extant co-design literature.

## Background

Co-design has its roots in various research and design traditions and disciplines. One of the most influential sources is the Scandinavian political participatory design tradition, which originally focused on workplace democracy and on worker and union empowerment in systems design. The tradition had a strong Marxist background with an emphasis on the conflict between capital and labour. Later, the tradition was criticised for losing its critical orientation and instead focusing too strongly on the participatory design process with associated design methods development. However, the approach has been revitalised during the past decades and the current participatory design tradition is again having a strong emphasis on empowerment – of various kinds of communities and groups suffering from marginalisation, oppression, or domination. In contemporary participatory design, participants' voices and views are underscored, equalising of power relations and democratic practices is emphasised, a highly reflective, ethical, and responsible stance is advocated, and mutual, reciprocal learning, competence building, and appreciation of each other's invaluable expertise are highlighted (e.g., Greenbaum & Loi, 2012; Luck, 2018; Pihkala & Karasti, 2016).

In addition to the management-hostile and critical approach of the Scandinavian participatory design tradition, criticality has also emerged in the field of design research as pictured in Critical Design (Dunne & Raby, 2001). Critical Design derives inspiration from the critical research tradition in social sciences and humanities, positioning itself in contrast to mainstream affirmative design and arguing design should be used to challenge the status quo (Dunne & Raby, 2001). Along these lines, critical design has been discussed in a broader sense and it has been argued that such critical design should foreground the ethics of design, reveal hidden agendas and values, and explore alternative agendas and values (Bardzell & Bardzell, 2013). Feminist design is among the approaches embodying these aims (e.g., Bardzell, 2010; Korsmeyer et al., 2022). Developments under the label of speculative design, in a similar way, argue for using design to open up possibilities, to debate, to consider alternative futures, to critique, to question, to provoke, and to raise awareness – potentially having a strong political role (Dunne & Raby, 2013). In design research inspired by critical and speculative design, various kinds of thought-provoking, speculative, critical, future-oriented artefacts have been created for these purposes.

Futures scholarship offers a related source of inspiration. According to this tradition, anticipating possible futures enables opening a discussion about people's fears and hopes regarding various futures (Amsler & Facer, 2017). A central idea among future scholars is that there is not a single future that can be predicted and that the future is not given but made in the present (Slaughter, 2018). Whyte et al. (2022) define such making as making sense of what might be possible and probable futures, engaging in a process of evaluation and negotiation to shape those preferred futures. From a critical perspective, the challenges go beyond anticipating and making futures, but encompass how to do so in order to foster alternative imaginations of worlds that escape the logic of oppression that characterises current systems. Visionary feminism may be well-equipped to tackle such a challenge since the cultivation of epistemologies and practices that contribute to building more equal and ethical worlds has been at the core of feminist imaginings and makings of future worlds (Coleman & Jungnickel, 2023).

Then again, drawing inspiration once more from the Scandinavian participatory design tradition, research on participatory infrastructuring has recently emerged, viewing infrastructuring as design for ongoing changes, ease of maintenance, and tailoring (Karasti & Baker, 2004), that is, design for "design-in-use" (Ehn, 2008). Participatory infrastructuring not only accounts for in-situ design activities but also highlights what takes place before, between, and after those activities as well as the participatory processes that take place vertically in different political and practical arenas and connect with existing networks and systems within and between organisations with agency being dispersed (Bødker et al., 2017). Participatory infrastructuring aims at sustainable changes in a longer time span – it calls for extended horizontal, vertical, and temporal reach, "stimulating the emergence of new knotworks through participatory activities in which participants actively engage in defining the infrastructures to be, and then sustain the networks of people and technologies that embrace the new artefact(s)" (Bødker et al., 2017).

## Methods

This chapter reports on design research, that is, research that entails changing the reality within which it operates through design. Hence, design research is generative; it is about a "proliferation of new realities" (Gaver, 2012). Design research entails the generation of design artefacts embedding researchers' (implicit) theories which are afterwards explored in use (Gaver, 2012). This design research study has explored ways for co-cultivating co-design methods for the purpose of promotion of gender equality in HEIs, aiming to integrate participatory and critical aspects into the co-design method repertoire. We report on three studies, all qualitative in nature:

- **Study A** Co-design of gender impact assessment (GIA): this study explored the futures of GIA inspired by research on co-design, future-making, and anticipatory methods.
- **Study B** Co-design of work–life balance: this study explored and raised awareness on parenting and work–life balance challenges inspired by co-design, critical, and speculative design methods.
- **Study C** Co-cultivation of co-design: this study explored the potential, contextual aspects, and sustainability of co-design in HEIs, inspired by research on participatory design and infrastructuring, with two rounds of interviews.

### *Participants*

In Study A, 11 participants representing Technology, Social Sciences and Humanities, and Life Sciences took part. All participants worked at the University of Oulu and had diverse experience levels and positions, and included professors, lecturers, postdoctoral researchers, and research service specialists.

In Study B, a group of seven participants was recruited from the University of Oulu. All participants had children, the participants represented different genders, disciplines, and functions in the university.

In Study C, during the first round of interviews, there were 13 participants from five HEIs representing five European countries, with a strong background in gender studies and/or work on gender equality. During the second round, two sessions in two HEIs representing two European countries were organised with eight participants who all had a strong background in gender studies and/or work on gender equality.

### *Co-design process*

In Study A, the theoretical underpinning lies in the understanding of co-design as a futures-making process in which many possible futures are anticipated (Akach et al., 2021; Sanders & Stappers, 2014), and that entails a collective negotiation of diverse values (DiSalvo, 2010; Van Mechelen et al., 2017). A key concern was how to support a space where participants would feel comfortable to express their views about the sex and gender dimension in research and thus, genuinely participate in the design of the GIA checklist. Puppet Scenarios method (Kumar, 2012) was considered a valuable format for creating a safe space to express one's thoughts, emotions, and even scepticism regarding the EC's new mandatory requirement for research applications (EC, n.d.). The Puppet Scenarios method was adapted and framed as a futures-making and anticipation activity. In the sessions, participants were encouraged to engage in playful explorations of possible futures by creating fictional characters (personas) and situations portraying how these characters related to the GIA checklist (scenarios). The participants created quick-and-dirty puppets representing their personas using the craft materials available at the sessions such as paper bags, colour papers, foam balls, ice-cream sticks, and glue. The stories described the personas' encounters with the GIA checklist, resulting in various outcomes. These stories were enacted by the participants through the puppets they created during the session. Each performance was followed by a group discussion on the issues raised in the stories.

In Study B, the aim was to examine and raise awareness and debate in HEIs on parenting, care, and work–life balance. Inspiration was derived from participatory design in the sense of inviting participants to design for better support for parenting, care, and work–life balance – something they all cared about and were experiencing conflicts with, but also from critical and speculative design traditions suited to raising debate and awareness through critique and provocation (Bardzell & Bardzell, 2013; Dunne & Raby, 2013). The cultural probes (Gaver et al., 1999) method was considered suitable for offering rich insights into the participants' everyday lives from the perspective of situations when family/parenting became visible/intertwined with work, problems in combining family life/parenting and work, and how family-related issues were represented in the HEI. The participants kept a diary for one week, crafting their "daily schedule" from this perspective. The participants also wanted to report on the benefits of combining parenting/family life and work, which was added as an aspect to the study. The participants were invited to creatively use text, drawings, video clips, and photographs in the study. Following the cultural probes study, the participants shared their insights in a workshop. Afterwards, another workshop was organised in which the participants were encouraged to consider their shared results as well as the message they would like to deliver in a media campaign aimed at raising awareness and debate on parenting, care, and work–life balance in HEIs.

Study C entailed reflective workshop discussions on co-design early and late in RESET. Early, the goal was to discuss initial insights on the potential of co-design for the promotion of gender equality in different cultural settings, while during the later phase, the goal was to share the gained experiences of co-design. For the initial sessions, inspiration was derived from participatory design; the goal was to reflect with the participants on the political, participatory, critical, and empowering underpinnings of co-design for gender equality. For the latter sessions, inspiration was also drawn from participatory infrastructuring that argues for design-in-use, for a consideration of maintenance and tailoring in use, and for broad horizontal and vertical participation for sustainable change (Bødker et al., 2017; Ehn, 2008; Karasti & Baker, 2004).

### *Data collection*

In Study A, the data consists of video and audio recordings, observations, and artefacts crafted by participants during three co-design workshop sessions in which the participants created and enacted stories (n=11), consisting of a persona and a scenario portraying possible ways how a tool for supporting GIA (the GIA checklist) might be used in the future.

In Study B, the data was collected through a week-long cultural probes study, including participant-created diaries with text and photographs. In addition, the data was collected through observation and note-taking during two participatory workshop sessions on data analysis and media campaign material development. The media campaign material collaboratively created by the participants is also included in the analyses.

In Study C, the data was collected through video recording of seven semi-structured group discussions entailing collaborative reflection on co-design, each lasting 1–2 hours.

### *Data analysis*

In Study A, the data collected was analysed by the research team generating insights to inform further iterations of the GIA checklist, with the workshop participants also being invited to analyse the workshop stories. Three out of 11 accepted the invitation. We collectively examined the research data, paying attention to the participants' explicit and tacit views on key aspects connected to the GIA checklist. We first identified the work practices and embodied aspects present and privileged in the stories, after which we explored not only what was said directly but also what was implied in the stories (see Durall et al., 2023).

In Study B, the data analysis was participatory, iterative, and data-driven in nature. Between the two participatory workshops, an initial data-driven analysis on the key issues and challenges relating to family/parenting becoming visible/intertwined with work was carried out by one researcher, combining and contrasting the findings from the individual cultural probes studies. The results were shared as a basis for the second workshop, in which the participants continued the analysis. As a result, a list of prioritised work–life balance and parenting issues and challenges was created. In addition, the participants came up with prioritised messages for the media campaign, with ideas for three thought-provoking media campaign artefacts (two blog posts and one video), based on which the artefacts were later developed by the participants and the researchers.

In Study C, the data analysis was iterative and data-driven in nature, entailing a careful examination of the workshops and reporting of the insights particular to each, as well as combining and contrasting the findings with other sessions. As a result of analysis of the first five sessions, a categorisation of local circumstances, discourses on gender equality, and sources of friction to be acknowledged in co-design for gender equality in HEIs emerged (see Iivari et al., 2023), with prior literature acting as a sensitising device helping to structure the empirical findings. For the last two sessions, a data-driven analysis was carried out to identify the ways of, and needs for, co-cultivation of co-design for sustained change in the HEIs.

## Empirical insights

### *Co-design of GIA tools*

The personas created by the participants represented university staff conducting research and research services specialists. In most of the scenarios (n=9), the GIA checklist was described as a support tool that the individual characters either benefited from or struggled with, when preparing research applications. In only two cases, the focus was on community impact, either by pointing to the organisation's power dynamic affecting the distribution of work or by highlighting how tools such as the GIA checklist might contribute to building more diverse and inclusive communities.

The stories created by the participants anticipated possible futures of addressing the gender dimension in research using the GIA checklist. The stories pointed to positive and negative impacts, but also presented complex situations in which diverse interests and agendas intertwined based on roles and power relations. While the stories were presented as tales about the future using a naïve and humorous tone, a closer examination revealed various levels of interpretation.

In their stories, the participants acknowledged that attention to GIA leads to better research, and explicitly mentioned several aspects characterising research practices, such as the competition to acquire research funding, the coordinator's individual effort when preparing research applications, and the late consideration of the sex and gender dimension when drafting the applications. The participants were also vocal about their imagined personas' gender, the consideration of sex and gender dimension in research as embodied, and the researchers' responsibility to address this dimension when planning research. The stories also included details about the emotions experienced by the personas in the GIA checklist use scenarios.

A second level of analysis focusing on the implicit meanings suggested the assumption that change in research practices requires a top-down approach, and that the short-term anticipated impact of addressing the sex and gender dimension in research may increase paperwork, while not necessarily leading to better research. The stories also tackled key issues affecting researchers, such as their capability to control their work, well-being in academia, and perceptions of the sex and gender dimension in research as another problem to address in order to obtain funding.

While the number of participants involved in the workshops was small (n=11), the stories and the following discussions held during the sessions showcased very diverse views on a topic (sex and gender dimension) that is controversial. Rather than reaching consensus (as mainstream co-design actions would aim for, in order to validate design decisions), our adaptation of the Puppet Scenarios method allowed participants to express diverse and sometimes clashing wishes and fears generated by the new EC requirement, and their expectations regarding a tool such as the GIA checklist. The analysis of the workshop outcomes pointed to various strategies such as the role-playing of personas as a mask from where to voice challenging views without feeling judged or creating conflict, as well as using humour to critique and question the status quo in academic research.

### *Co-design for work–life balance*

Based on the cultural probes study, the participants discussed sensitive topics in a highly open and honest manner, identifying common and significant themes from the data, mostly pain points but also some occurrences where parenting/family impacted work in a positive manner.

Afterwards, the group considered the message they would like to deliver in the media campaign. This work resulted in two blog posts and a video. One blog post, called "Life in the Academia: Caring for Sick Children and Feeling Guilty", points out that children get sick often, which seriously affects the participants' work. The post maintains that flexibility in the work is appreciated; however, it also creates a serious burden and makes the participants feel guilty: "Due to taking care of my sick children, I had to postpone two tasks until after they fell asleep. As a result, I had to compensate for my working hours from 9 pm to 11:30 pm".

> My son is having a flu and cannot go to day-care. I have a review deadline today. So, it requires arrangements. . . . I am sick too, and very tired. Around two my husband comes home, and I get to work for a couple of hours. At night, I'm too tired to work. And feeling guilty of having such a bad workday.

Another blog post, "Life in the Academia: Research Mobility – a Blessing and a Curse" directs attention to the challenges involved with research mobility, which is expected and among the benefits of the work:

> I will be embarking on a research visit. My family will be accompanying me, which is a fantastic opportunity. This arrangement allows my family to share my experience, and we won't have to live apart during this time. This is a unique benefit that many other professions may not offer.

However, the family may not always be able to join the trips and the burden for the family as well as feeling bad for not being with them is brought up:

> My partner, who is a researcher, travels quite a bit and these times are especially taxing, as home duties cannot be distributed equally. Children wake up. Partner, who normally handles the morning shift, is on a work trip. As a non-morning person, this affects coping with the work and chores.
>
> Leaving my son to daycare crying . . . It feels like he is reacting this way due to me being away for the previous week. He says that he wants to stay home with me and doesn't want to go to daycare. I know this feeling will pass and he will be just fine, but it doesn't feel good.

The post ends with a call for enhanced university support for parenting academics on research visits.

The group also decided to create a video, the creation of which turned out to be a highly organic process with new enthusiastic participants joining in. Five new participants and experts in video production joined the work and a workshop for ideation was organised to produce a video with the purpose of raising reflection among the viewers about the challenging combination of work and family life, but also about the counterbalance offered by family life and the joy that family brings amid everyday work life. The participants recorded children's voices, discussions, and thoughts in casual settings. They also recorded messages received from home, spouse, and public institutions, such as kindergarten, school, and hobby groups. A film producer and a producer from the group combined a soundscape with the video. The video tells the story of a tired researcher whose everyday life is filled with messages coming from different directions. The video consists of audio and video clips and a researcher staring at the screen, burdened by the flood of messages arriving from different directions. The video ends with a counterbalancing atmosphere, where the researcher's home life is described with homecoming greetings, hugs, and happy thoughts that arise when the researcher meets his family after a working day.

Overall, during this process, many of the participants worked enthusiastically and shared personal stories of their life in an open manner. We consider that the relatively homogenous group of participants – all sharing similar experiences of parenting and work–life balance challenges – and the cultural probes study that invited them to share personal everyday life stories led to the creation of an atmosphere in which the participants felt safe to discuss the pain points in a very open and honest manner. The invitation to raise awareness and debate on these challenges was also appreciated. The creation of the video in particular sparked a lot of engagement.

### *Co-cultivation of co-design*

The initial sessions aimed at creating a shared understanding of co-design, its potential, and the contextual factors shaping it. The findings reveal a variety of contextual factors and their historical trajectories shaping co-design for gender equality in HEIs in different European countries. At the national level those included national politics, legislation, religion, and public discourses, while at the organisational and disciplinary levels, hierarchical and political factors inside the HEIs and various ideologies and discourses on gender equality had clear impacts. A lot of friction to be noted or dealt with was identified. The findings underscore how political and power-related co-design is in the context of HEIs and in the political milieu of European countries. The findings also underscored how vital it is to be positioned on the side of the marginalised to combat their oppression and strive towards democratic practices.

The latter sessions aimed at sharing co-design experiences and ensuring sustained changes in the HEIs. The experiences concern the ways by which the RESET members had been engaged in co-design for gender equality in RESET, but also how the HEI staff and students had been engaged in co-design and thus contributing to gender equality work through different kinds of events, methods, organisational bodies, and positions. The findings show new methods and tools had been experimented with, thanked for enabling sharing (personal and sensitive) experiences, engaging the participants and equalising power relations, and resulting in concrete outcomes. Co-design with broad participation, especially in collecting feedback, had taken place, sometimes even without RESET members' initiative. The inclusion of some previously marginalised stakeholders was mentioned. The participants pointed out eagerness and enthusiasm but also resistance and passivity inside their institution, reminding about the cultural sensitivity of co-design in the academic context. The participants discussed how people in some positions and bodies had already taken responsibility for the promotion of gender equality in their institution, but also problems in this respect that they wanted to be discussed in future-oriented sessions, which aimed to invite the local HEI representatives to discuss the future of gender equality in their organisation. These sessions were to include the presentation of experiences and practices of University of Oulu as a basis for the discussion as well as collaborative discussion of the strengths, weaknesses, values, and marginalised stakeholders in the HEIs in question as a basis for brainstorming and visioning of the future (inspired by Beyer & Holtzblatt, 1998; Dunne & Raby, 2013; Kumar, 2012), resulting in a concrete action plan. In these sessions, participants representing gender equality work and participants representing higher level decision-making positions were to participate, with the goal of raising awareness and action-taking for sustained change and continuation of the work after RESET.

## Concluding discussion

With this chapter, we have made visible the work we have done in RESET on co-cultivating co-design, in collaboration with the RESET members but also with students and staff of the involved HEIs. We maintain that for co-design of gender equality practices, tools, and policies, mainstream co-design should be cultivated with participatory and critical aspects, which can also be seen to underline the roots of the co-design approach.

Strongly inspired by the Scandinavian political participatory design, we have advocated for encouraging broad participation in all co-design sessions, specifically considering previously marginalised participants: giving a voice and a say to them through democratic practices. In addition, inspired by the developments in critical and speculative design and futures-making and anticipation traditions, we have underscored the use of design for critique, for questioning the status quo, for exploring alternative futures, agendas, and values, and for raising debate and awareness.

We think that these traditions suit co-design for gender equality very well. We argue that participatory design and gender equality work can be considered kindred spirits: both prioritise equality, ethics, reflexivity, and positionality and work on behalf of those suffering injustice – hence, their values are aligned (see Iivari et al., 2023). Due to the political aims of gender equality work, the critical and speculative design traditions also serve it admirably: both aim to critique the status quo, question, and advocate for political change. Similarly, futures-making and anticipation traditions invite for anticipating, imagining, and making of more equal and ethical futures (Amsler & Facer, 2017; Coleman & Jungnickel, 2023; Slaughter, 2018; Whyte et al., 2022). There has already been cross-fertilisation between these traditions: there are discussions, for example, on feminist design, futures-making, and anticipation (Bardzell, 2010; Korsmeyer et al., 2022; Whyte et al., 2022). We wish to make the shared value base and aims even more visible, to push mainstream co-design towards better serving gender equality.

We also wish to highlight the contribution of participatory infrastructuring, which makes it evident that for social or political change to be maintained, we should design for sustained use, for maintenance and tailoring in use, and for participatory processes with great horizontal and vertical reach. In RESET, we have emphasised broad horizontal participation throughout the years, while, during the last year, we have placed increased emphasis on vertical reach with participants representing decision-making positions.

Based on our empirical and theoretical work, we propose the following principles for co-design in the context of gender equality work in HEIs. We recommend **1) modifying co-design to provide a safe place to explore and articulate controversies and power-laden topics** – this is often needed in the case of gender-equality work. Another recommendation relates to **2) modifying co-design for raising critique, questioning, and exploring alternative futures, values, and agendas** – for gender equality, this is certainly needed, that is, critique for social and political change. In addition, we call for **3) fostering broad and inclusive participation, in the best case that which organically emerges and evolves in the HEIs due to participants' own interest and initiative**. Finally, we call for **4) participatory infrastructuring for sustained social change**, which includes designing for the

**TABLE 5.1** Emerging principles for co-design for gender equality in HEIs

| | *Co-design of GIA tools* | *Co-design for work–life balance* | *Co-cultivating co-design* |
|---|---|---|---|
| **Theoretical background** | Co-design inspired by futures-making and anticipatory practices | Co-design inspired by participatory, critical, and speculative design | Co-design inspired by participatory design and infrastructuring |
| **Practical methods** | Puppet theatre; joint data analysis | Cultural probes; joint data analysis; joint design of media campaign artefacts | Reflective workshop discussions; brainstorming; and visioning |
| **Principles for co-design for gender equality** | Safe place to explore and articulate controversies and power-laden topics;<br>Design as critique, design of thought-provoking artefacts for raising awareness and debate | Safe place to explore and articulate controversies and power-laden topics;<br>Design as critique, design of thought-provoking artefacts for raising awareness and debate;<br>Emergent enthusiastic participation | Safe place to explore and articulate controversies and power-laden topics;<br>Design for design-in-use, sustained use, maintenance, and tailoring;<br>Broad participation with extended horizontal and vertical reach |

emergence of networks and communities that take ownership and responsibility for the maintenance, update, and sustainability of gender equality practices, tools, and policies. Table 5.1 links these principles with the studies conducted with associated theoretical and methodological underpinnings.

There are several limitations and paths for future work to be noted. The study was carried out in the RESET project with particular HEIs representing particular European countries. Many sessions included only a limited number of participants from the University of Oulu. In the future, the methods we explored should be experimented with in other cultural settings with broader participant groups. We also think that the speculative and critical design traditions could be integrated even more strongly into the co-design method repertoire to inspire thought-provoking designs that aim to trigger reflection and critique and to push towards increased gender equality. Another aspect to be further explored concerns ways by which to conduct participatory infrastructuring for gender equality in HEIs: although this is something we worked for in RESET, we could have utilised this notion and its methodological underpinnings earlier and more extensively for inviting local HEI communities and decision-makers to develop their local visions and to take ownership of and responsibility for the evolving gender equality practices, tools, and policies.

## References


Akach, Joyce Akinyi, Osanjo, Lilac, & Maina, Samuel. (2021). Co-design: Tools, methods and techniques for designing with users. *Africa Habitat Review Journal*, *15*(1), 2145–2156.

Amsler, Sarah, & Facer, Keri. (2017). Contesting anticipatory regimes in education: Exploring alternative educational orientations to the future. *Futures*, *94*, 6–14.

Bardzell, Jeffrey, & Bardzell, Shaowen. (2013). What is "critical" about critical design? In *Proceedings of the SIGCHI conference on human factors in computing systems*, 3297–3306. New York, USA: ACM.

Bardzell, Shaowen. (2010, April). Feminist HCI: Taking stock and outlining an agenda for design. In *Proceedings of the SIGCHI conference on human factors in computing systems*, 1301–1310.

Beyer, Hugh, & Holtzblatt, Karen. (1998). *Contextual design: defining customer-centered systems*. Elsevier: San Diego.

Bødker, Susanne, Dindler, Christian, & Iversen, Ole S. (2017). *Tying knots: Participatory infrastructuring at work. Computer Supported Cooperative Work*, *26*, 245–273.

Coleman, Rebecca, & Jungnickel, Kat. (2023). Introduction to creating feminist futures: Research methodologies for new times. *Australian Feminist Studies*, 1–13.

DiSalvo, Carlo. (2010, July 7–9). Design, democracy and Agonistic pluralism. In David Durling, Lin-Lin Chen, Tiiu Poldma, Seymour Roworth-Stokes, & Erik Stolterman (Eds.), *Design and complexity – DRS international conference*. Montreal, Canada: Design Research Society.

Dunne, Anthony, & Raby, Fiona. (2001). *Design noir: The secret life of electronic objects*. Springer Science & Business Media.

Dunne, Anthony, & Raby, Fiona. (2013). *Speculative everything: Design, fiction, and social dreaming*. MIT Press.

Durall, Eva, Iivari, Netta, Heikkinen, Mervi, Pihkala, Suvi-Tuulia, Kinnula, Marianne. (2023, June 12–14). Anticipating the futures of the gender dimension in research: Storying entangled practices and bodies. In Vanessa Rodrigues, Carl Westin, & Stefan Holmlid (Eds.), *Nordes 2023: This space intentionally left blank*. Norrköping, Sweden: Linköping University.

Ehn, Pelle. (2008). Participation in design things. In *Participatory design conference*, Bloomington, Indiana, USA, 92–101.

European Commission. (n.d.). *Gender equality in research and innovation*. https://research-and-innovation.ec.europa.eu/strategy/strategy-2020–2024/democracy-and-rights/gender-equality-research-and-innovation_en

European Commission, Directorate-General for Research and Innovation. (2021). *She Figures 2021, Gender in research and innovation – Statistics and indicators*. Publications Office, https://data.europa.eu/doi/10.2777/06090

Gaver, Bill, Dunne, Tony, & Pacenti, Elena. (1999). Design: Cultural probes. *Interactions*, *6*(1), 21–29.

Gaver, William. (2012). What should we expect from research through design? In *Proceedings of the SIGCHI conference on human factors in computing systems*, 937–946. New York: ACM.

Greenbaum, Joan, & Kyng, Morten. (Eds.). (1991). *Design at work: Cooperative design of computer systems*. Lawrence Erlbaum Associates Inc.

Greenbaum, Joan, & Loi, Daria. (2012). Participation, the camel and the elephant of design: An introduction. *CoDesign*, *8*(2–3).

Iivari, Netta, Sharma, Sumita, Ventä-Olkkonen, Leena, Molin-Juustila, Tonja, Kuutti, Kari, Holappa, Jenni, & Kinnunen, Essi. (2022). Critical agenda driving child–computer interaction research – taking a stock of the past and envisioning the future. *International Journal of Child-Computer Interaction*, *32*, 100408.

Iivari, Netta, Tervo, Erkki, Käsmä, Marjukka, & Heikkinen, Mervi. (2023). Participatory design meets gender equality at European higher education institutions. *CoDesign, 19*(4), 304–326.

Karasti, Helena, & Baker, Karen S. (2004). Infrastructuring for the long-term: Ecological information management. In *Proceedings of the 37th annual Hawaii international conference on system sciences.*

Korsmeyer, Hannah, Light, Ann, & Grocott, Lisa. (2022). Understanding feminist anticipation through 'back-talk': 3 narratives of willful, deviant, and care-full co-design practices. *Futures, 136*, 102874.

Kumar, Vijay. (2012). *101 design methods: A structured approach for driving innovation in your organization.* John Wiley & Sons, Inc.

Luck, Rachael. (2018). What is it that makes participation in design participatory design? *Design studies, 59*, 1–8.

Pihkala, Suvi, & Karasti, Helena. (2016). Reflexive engagement: Enacting reflexivity in design and for "participation in plural". In *Proceedings of the participatory design conference: Full papers* (Vol. 1, 21–30). New York, USA: ACM.

Sanders, Elizabeth B. N., & Stappers, Pieter Jan. (2008). Co-creation and the new landscapes of design. *Codesign, 4*(1), 5–18.

Sanders, Elizabeth B.-N., & Stappers, Pieter Jan. (2014). Probes, toolkits and prototypes: Three approaches to making in codesigning. *Codesign, 10*(1), 5–14.

Slaughter, Richard. (2018). Two fine additions to the futures literature. *Foresight, 20*(4), 443–446.

Van Mechelen, Maarten, Derboven, Jan, Laenen, Ann, Willems, Bert, Geerts, David, & Abeele, Vero Vanden. (2017). The GLID method: Moving from design features to underlying values in co-design. *International Journal of Human-Computer Studies, 97*, 116–128.

Whyte, Jennifer, Comi, Alice, & Mosca, Luigi. (2022). Making futures that matter: Future making, online working and organizing remotely. *Organization Theory, 3*(1).

# 6

# A NEW GENERATION OF GEPS

## How academic organisations address equality, intersectionality, and diversity

*Marisa Matias, Sara Isabel Magalhães, Jorge Peixoto Freitas, Alexandra Lopes, Conceição Nogueira, and Maria Isabel Dias*

### Introduction

This chapter examines the evolving landscape of Gender Equality (GE) initiatives within four academic institutions, focusing on the challenges, strategies, and successes in implementing GE Plans (GEPs). Gender Equality, meaning Equal rights, responsibilities and opportunities of women and men and girls and boys (EIGE, 2024), in higher education, is increasingly recognised not just as a matter of fairness but as a fundamental aspect of excellence and social justice. However, while progress has been made, the journey towards equality is far from straightforward. Universities and research institutions continue to face persistent gender disparities, both at the structural level and in everyday academic practice.

Drawing from the experience of the RESET project and its partners, this chapter will explore the complex dynamics involved in the design, implementation, and monitoring of GEPs. It highlights the need for a nuanced approach that considers not only gender but also other intersecting social categories such as age, professional group, socio-economic status, and sexual orientation. The concept of intersectionality is now becoming an essential framework for understanding and addressing the multi-layered nature of discrimination and privilege in academic settings. Through a critical examination of current practices, the chapter discusses how RESET universities tried to move beyond traditional gender-focused strategies to embrace more inclusive, intersectional approaches.

The chapter also reflects on the importance of co-design methodologies in developing effective GEPs. Such a participatory approach, with diverse stakeholders, ensures that the perspectives of all parties are considered and helps address resistance by fostering a sense of shared responsibility among stakeholders. Ultimately,

DOI: 10.4324/9781003564348-11

this chapter provides insights and reflections for higher education institutions (HEIs) seeking to advance GE in ways that are inclusive and context-sensitive.

## GEP as a tool in the European framework

GE is a concept that has been shaped by processes involving social, political, and institutional actors, as well as national and international organisations, influencing its meanings and GE policies (Lombardo et al., 2010). These meanings are multiple and sometimes more contradictory than coherent. Different visions of GE translate into different policy strategies (Walby, 2005). In consolidating an EU social policy framework, GE has become one prominent pillar and is one of the topics more often used as an example of what became known as the Europeanisation of national policies, understood here in a relatively simple way as the change in national policies as a result of pressures and/or developments in EU policies (Forest & Lombardo, 2012).

The European Union has been playing a leading role in proposing transversal policies, known as gender mainstreaming, which has more recently affected university organisations (Abels, 2011), in particular, since GEPs became a mandatory requirement to access European funding under the H2020 framework.

Despite the wide consensus about higher education and research playing a critical role in promoting social change, GE, diversity, and inclusion in society at large, the realisation that universities remain gendered and gendering organisations (Rosa et al., 2020) has shifted attention to promoting GE within academia as a necessary preliminary step. Mounting evidence about inequalities (i.e., unequal access to or enjoyment of rights, as well as the assumption of stereotyped social and cultural roles) such as persistent gender imbalances and pay gaps, gender segregation, both horizontally (across disciplines) and vertically (disproportional presence of women in higher levels of the academic hierarchies), as well as the public recognition of harassment and sexual assault as practices silenced over the years, have led to an acknowledgement of how academic institutions are structured on the basis of gender divisions of labour and power (European Commission, 2019; Hearn et al., 2022; Paoletti et al., 2020). It was in this *milieu* that the European Commission (EC) formalised its commitment to advancing GE in research and in higher education, with the Horizon 2020 framework representing a significant commitment from the Commission to foster GE. Horizon 2020 was shaped by three main dimensions of goals: closing the gap in participation, with an emphasis on promoting equal opportunities and gender balance in the composition of research teams; ensuring gender balance in decision-making processes, with an emphasis on composition of panels and advisory groups; and strengthening the gender dimension in research contents, with an emphasis on increasing the relevance of research for all persons and all groups in society.

One mechanism that has become increasingly nurtured by the EC in order to tackle gender inequalities in HEIs concerns the implementation of GE Plans

(GEPs). Under Horizon 2020, there were already specific calls for the implementation of GEPs, with funding allocated specifically for that purpose. This type of initiative excluded a significant number of universities from potential funding, and limited European countries that did not have national GE legislation or programmes from accessing funding. Thus, the only alternative for HEIs was to quickly design and implement a GEP.

GEPs are considered promising because they cover a very wide range of themes that are tackled taking into account the contextual and institutional factors of each academic organisation. GEPs cover recruitment and selection, and career development and progression, but also work–life balance, leadership and decision-making, the culture of the organisation, gender integration in research and education, as well as gender-related harassment, assault, and discrimination.

GEPs are required to be signed by top management and publicly advertised on the university website. Human resources with gender competence have to be allocated to the implementation of GEPs; sex/gender-disaggregated data have to be collected and monitored; annual reports based on indicators have to be published; and training for staff, including those in decision-making positions, has to be accounted for (EIGE, 2023). A GEP combines gender mainstreaming, that is, a systematic consideration of the differences between the conditions, situations, and needs of women and men (EIGE, 2024) with positive actions aimed at favouring access by members of certain categories of people, in this particular case, women, to rights which they are guaranteed to members of other categories (Bencivenga et al., 2021; Vida, 2020). GEPs are therefore a valuable tool for understanding gender inequality in HEIs and for making strategic changes in previously gender-insensitive universities (Bencivenga et al., 2021; Laoire et al., 2021).

To what extent GEPs are effective in triggering institutional and cultural changes in universities is far from being a consensual topic, with many criticising GEPs for resting on an overly instrumental rationale, while others highlight the conservatism of the frameworks that inform understandings of GE (Lätti, 2017). Yet, the dissemination of positive experiences in implementing GEPs in universities across Europe (Rosa et al., 2020) has definitely contributed to their wide acceptance as a promising practice, so much so that they would be resumed under Horizon Europe as an eligibility criterion.

## Key approaches to GEP design at RESET

All GEPs developed under the RESET project followed two key principles – co-design and intersectionality, both of which are detailed in other sections of this book. As current literature emphasises the growing importance of intersectionality in designing GEPs, particularly in research and innovation (R&I) sectors, applying intersectionality in GEPs requires structural changes rather than simply adding diverse categories to existing frameworks. For example, organisations are encouraged to audit career progression criteria to address biases and ensure fair representation in leadership roles. This aim comes up against the lack of data intersecting gender and other dimensions. Structural changes are essential to prevent intersectionality from being reduced to a buzzword. By embedding intersectionality deeply within GEPs, organisations can create genuinely inclusive environments that address the multifaceted challenges faced by individuals.

Another key approach in RESET GEPs is the use **of co-design**, that is, an emphasis on active and meaningful stakeholder participation (Iivari et al., 2023). This means identifying all relevant stakeholders who will use or be affected by the GEPs and inviting them to contribute to the design process. The process should encourage participants to learn from one another, recognising that combining diverse perspectives leads to creative and innovative solutions. As such, co-design approaches focus on equalising power relations and promoting democratic practices by actively including those who might otherwise be marginalised. In the RESET context, Gender Equality Boards (GEB) were a decisive aspect for co-design and for GEP's success. These boards have a steering role for Gender Equality Actions allowing for the support of governance to GEPs.

## GEP design at HEIs

Drafting an initial GEP from scratch can be daunting, as there is a considerable lack of information, and existing data are scattered and most likely not disaggregated by sex. First measures include approaches that set foundations for action. A first GEP cannot be a long list of measures. One has to be strategic to set the ones that are more relevant to be considered, bring more impact, and are contextual and data-driven. For this purpose, it is important to set the stage and know the *status quo* of the institution.

### *Setting the stage*

According to the GE in Academia and Research (GEAR) Tool, steps 1 and 2 of devising a GEP aim to get to know the institution in great detail to ensure the best adequacy and effective change of the proposed actions. One prior step we considered in our RESET approach was to get familiar with the national context and its developments in terms of gender and GE policies.

In this regard, the four institutions implementing the GE Plan within the RESET project come from four countries with significant differences in geography, language, and historical background, but also in the current sociocultural context: France, Greece, Poland, and Portugal. According to the most recent EIGE report[2] (European Institute for Gender Equality, 2024), in 2023, the countries of the four implementing partners stand far apart in terms of the GE Index. France is the only country above the EU average, ranking 6th, Portugal ranks 15th, Poland 18th, and in the last position of this index is Greece, ranking 25th.

While all four universities are committed to promoting gender equality, their contexts and approaches vary. AUTh faces significant challenges in gender equality, with no women in its rectorate body and a low percentage of women in senior administrative positions. Despite these challenges, Greece has made efforts to align its legal framework with EU policies on gender equality. Laws mandate gender representation in decision-making bodies and promote gender equality in research and higher education. The National Action Plan for Gender Equality 2021–2025 included actions to promote gender equality in education, science, research, and leadership positions. The University of Bordeaux has implemented numerous measures to promote gender equality, including the Masterplan for Gender Equality (2015–2020) and the Action Plan for Professional Equality between Women and Men (2021). The university has a dedicated GEB and has expanded the remit of the Sexual Harassment and Sexist and Homophobic Violence Monitoring Unit. The university's commitment to GE is recognised with the "Sustainable Development and Societal Responsibility" label. Poland has faced anti-feminist backlash and resistance to GE regulations, creating challenges for the implementation of GEPs, nevertheless at UL, gender equality and inclusivity have been prioritised, with the mission to conduct reliable research and educate future generations. The majority of UL employees are women, and the university recognises issues faced by trans- and gender-diverse persons. Finally, at the University of Porto, there is a long-standing commitment to GE, reflected in its Value Statement and Strategic Plan. The university conducted its first gender-impact assessment in 2020 and has implemented a directive on work–life balance. Nevertheless, women are also lacking in top management positions and these core values have been difficult to incorporate in effective practices. As a member of the European University Alliance for Global Health (EUGLOH), the university is actively involved in promoting gender equality and diversity inclusion.

Knowing this national context was important as it shaped the academic environment and is a necessary step to a proper diagnosis of the GE situation. This diagnosis encompassed a combination of different sources of data, including the use of secondary sources and available information within university servers, collecting HEI worker's experiences and perceptions through a survey and through focus groups. Data was collected according to four priority areas, recommended in the GEAR tool:

- Recruitment, retention, and career progression including the availability of family-friendly policies;
- Leadership and decision-making (accountability, transparency, and inclusiveness);
- Gender dimension in research and knowledge transfer (content and curricula);
- Gender biases and stereotypes, sexism, and sexual harassment.

Data collection from an intersectional perspective comes up against several obstacles. The first was encountered as the information was dispersed across the universities: not only was there no centralised source of information but also sex-disaggregated data was not always available. Moreover, other socio-demographic information, such as ethnicity and disabilities of HEI's workforce, was unavailable, limiting the intersectional approach.

While the secondary data gave a general overview of the university, the focus groups, performed with diverse key audiences – teachers, researchers, staff members, and top leadership – were able to complement the absent data and provide insights on how the HEI's staff was experiencing these issues within the university (see more on this data in Chapter 7 of this book).

Lastly, a large-scale survey collected information on perceptions and experiences of GE, biases, and resistances. The major obstacles here related to the low response rate and again to the difficulties in performing an intersectional analysis due to the preservation of participants' anonymity and non-disclosure of identifying information.

### *Crafting a first GE plan*

Considering the academic context of each partner and the insights and findings deriving from the data collection, an extensive diagnosis was performed for each partner. Due to the complexity and the specificity of the respective contexts, a subset of the most salient findings and respective measures from each local GEP will be reflected here. **Horizontal segregation and lack of recruitment strategies to prevent biases were described by all partners when addressing the state of play in their institutions.** Women are the majority among employees in RESET institutions, particularly in lower status categories and in traditional fields of knowledge. They are less represented at higher hierarchical levels, except in areas where they are already a majority. Also, in all institutions, motherhood significantly impacts women's career progression, leading to slower advancement compared to men.

All these markers of inequality align with known processes within the literature, such as the "glass ceiling" metaphor. Invisible barriers exist, as formal rules do not put any limitation on women progression; however, biases in how the recruitment and promotion processes are undertaken contribute to this issue (Clavero & Galligan, 2021). The tip of the iceberg comes from the under-representation of women among professors and university rectors. Transversal measures that partners endorse in their GEPs are linked to reviewing and rewriting existing recruitment guidelines to make them more gender sensitive and to raise awareness of and combat gender stereotypes that are associated with scientific areas. At U.Porto, actions encompassed reviewing the process manual of recruitment and at UBx, the selection committees were targeted at raising awareness actions on gender stereotypes in recruitment. At UBx, a serious game, titled *Catch Me If You Can!*, was developed with this topic in mind. Another important stream of action included communication campaigns, such as those developed in UL, U.Porto, and AUTh, that made researchers of the under-represented gender visible in all research fields, also combating gender stereotypes regarding academic career development. At UL, the issue of women's under-representation in managerial positions was further tackled by foreseeing a revision of the eligibility criteria for promotions. This was crucial, as the criteria used to chart career progression often do not encompass diverse pathways characterised by a discontinuous investment in scientific or pedagogical areas.

An ambiguous **discourse about quotas** was also widespread across partners' diagnoses. While quotas were recognised as tools for structural change, they were seen as conflicting with the notion of meritocracy. This tension was highlighted by participants describing "imposter syndrome" where individuals hired or promoted under quotas feel undervalued. Indeed, when gender quota systems are poorly implemented, it can undermine women's perceived value and companies' value within the market. Thus, while quotas are important, they alone cannot address gender inequalities (Rixom et al., 2023). At U.Porto, quotas were introduced on boards that make the decisions instead of introducing them to the application/selection process itself, as this would entail a lot of resistance; therefore, selection panels for career progression competitions need to have a 40/60 gender balance.

The issue of **parenthood's impact on women's career is interrelated with career progression and the struggle to reconcile professional and personal lives.** In the diagnoses, women were found to face more hardships in terms of balancing work and family. Leave-taking and the use of family-friendly policies were also highly gendered, with women using them most often. Men, meanwhile, reported prejudice when taking on caregiving roles. Most academics work evenings and weekends to meet diverse demands, such as innovative research, publishing in high-impact journals, teaching, supervising students, building networks, and securing research funding. Furthermore, excellence is linked to mobility, requiring academics to spend time at different universities, often abroad. Women face greater normative expectations for household and caregiving responsibilities (Rosa, 2022). In addition, work–life policies have limited effectiveness as traditional cultural norms about the gender division of labour and the ideal male worker persist in organisational culture. Although policies confer equal formal rights, informal expectations often pressure women to use these rights, while men remain focused on academic activities. These issues are also the priority areas for all partners' GEPs. At UL, a sensitive and flexible system for evaluating academic performance was designed, and UBx implemented a rotation in teaching and research responsibilities so that teachers and researchers have an opportunity to dedicate more time for research or teaching depending on their own choice. Measures such as training managers to better advise their teams on work–life balance measures (UBx), implementing a room for lactation in every school (AUTh), and an online portal encompassing work–life balance rights and measures for all HEI's staff (U.Porto) were also implemented.

With regard to **gender mainstreaming within the core activities of academia and research and knowledge transfer**, no formal policies or incentives exist. Female teachers and researchers often rely on informal practices (e.g., using some examples in classes, inviting male and female lecturers). While Horizon Europe funding encourages gender mainstreaming, its added value remains debated. In all RESET GEPs, actions devoted to capacity-building and awareness-raising were developed. Researchers, professors, lab and faculty directors, and managers were trained on gender inequalities and possible bias in teaching and researching. These actions were also accompanied by providing specific tools developed within RESET – the Gender Impact Assessment Checklist to support the integration of gender dimension in academia.

Throughout the diagnoses, **intersectional discrimination was evident,** that is, discrimination intersecting gender and age (mostly) but also gender with professional groups (research, administrative staff, and teachers). A common finding across partners was the power asymmetries between administrative staff and teachers/researchers, driven by the hierarchical and elitist structures of HEIs. This perceived status difference may also hinder a collective sense of excellence as it emphasises competition based on individual performance. This challenge was approached by including all HEIs employees in capacity-building sessions and by applying work–life balance initiatives to all. Other reported sources of discrimination include sexual orientation and gender identity, ethnicity/nationality, or religion reminding us that power imbalances are multidimensional. In most partners, reporting mechanisms for gender-based violence, sexism, and sexual harassment were lacking. Despite this, harassment cases are widely acknowledged but often go unreported or uninvestigated ("black figures"). In addition, Gender-Inclusive Language (GIL) was also found to be a source of resistance and a systematic use was clearly absent. To tackle this challenge, the promotion of the use of gender-inclusive language supported by the RESET toolbox for gender-neutral, diversity-oriented institutional communication was included in all GEPs, with some partners adapting the toolbox to its own language (e.g., U.Porto). Another pack of measures covered the introduction or refinement of existing procedures to ensure a secure reporting system of cases of discrimination, sexism, or harassment. U.Porto launched a secure reporting mechanism for discrimination and harassment situations and put forward a commission to assess and screen the complaints. This procedure was accompanied by an update on the employees' and students' code of conduct and by training sessions. At UBx, actions were devoted to raising awareness on the existing system and on improving it. At all partner institutions, another set of training and capacity-building actions was focused on abilities to identify and report discrimination, violence, or harassment situations. One major action that oversees all of these is an explicit statement by the governance of a zero-tolerance policy and a code of conduct against discrimination and moral and sexual harassment.

The two key principles of RESET – co-design and intersectionality – underlined both the development process and contents of the GEPs. The co-design approach was carried within each partner by cooperative work with multiple stakeholders, ensuring the engagement of all and the commitment of senior leaders. A key step in this process was establishing GEBs at each university early on GEP design, and involving them from the data collection stage. These Boards' composition was adapted to each partner alongside two dimensions: interest/commitment to gender equality and power within the institution. As such, a diverse group of members were part of this GEB. All GEBs were mandated by the Rector of each university with the mission to support and contribute to the design and monitoring of the GEP. However, not all were provided with dedicated time or budgetary resources. To design the GEP, besides GEB's inputs, RESET teams met with senior management (e.g., vice-rectors or chancellors) and key institutional offices. This inclusive approach fostered both a top-down and bottom-up process, mitigating challenges and resistances to GEP implementation.

Figure 6.1 illustrates the whole process for the GEP development, from step 1 of the GEAR tool[1] – Getting Started – to step 5 – Monitoring Progress and Evaluating a GEP. This scheme also highlights the co-design approach followed throughout this process.

### *Paving the way: implementing and monitoring*

Having set up a GEP is the tip of the iceberg of all the work to be done, accompanying the implementation up close and constant monitoring is crucial.

RESET GEPs include actions for capacity-building, improvement of data monitoring systems, revising key documents to integrate GE and diversity, and establishing mechanisms for reporting discrimination. GEPs aimed both at a structural change with actions devoted to regulations, mechanisms for reporting discrimination, and changes in recruitment and progression procedures, but also emphasised individual change through actions such as training, and the promotion of inclusive language and of gender impact assessment tools in the daily life of HEI members.

To assist in the implementation stage, some partners have dedicated GE structures (e.g., AUTh Committee) or missions (e.g., UBx's), while others lack such facilities. In addition, as so nicely put by Roggeband and Verloo (2006), one of the main problems in the implementation of gender mainstreaming is that it paradoxically requires a high level of gender awareness among policymakers who are not gender experts. Thus, the involvement of key local stakeholders that can, effectively, implement the measures outlined in the GEP and transversally mainstream GE in the community and their practices is a key factor for successful implementation and for GE sustainability. In RESET, GEBs were the steering structures for GEP implementation and for GEP sustainability. These structures accompanied the entire process of GEP design and implementation, while the GEP implementation team primarily consisted of RESET team members and selected representatives from university services.

All partners' GEPs aimed to institutionalise this board beyond the RESET project to ensure the sustainability of the changes initiated. At AUTh and UBx, such a board is mandated by national law, and thus some resources (such as time allocation is already guaranteed). At UL, institutionalisation took place early on, and with the recent elections, the area of Equal Treatment and Non-Discrimination is now

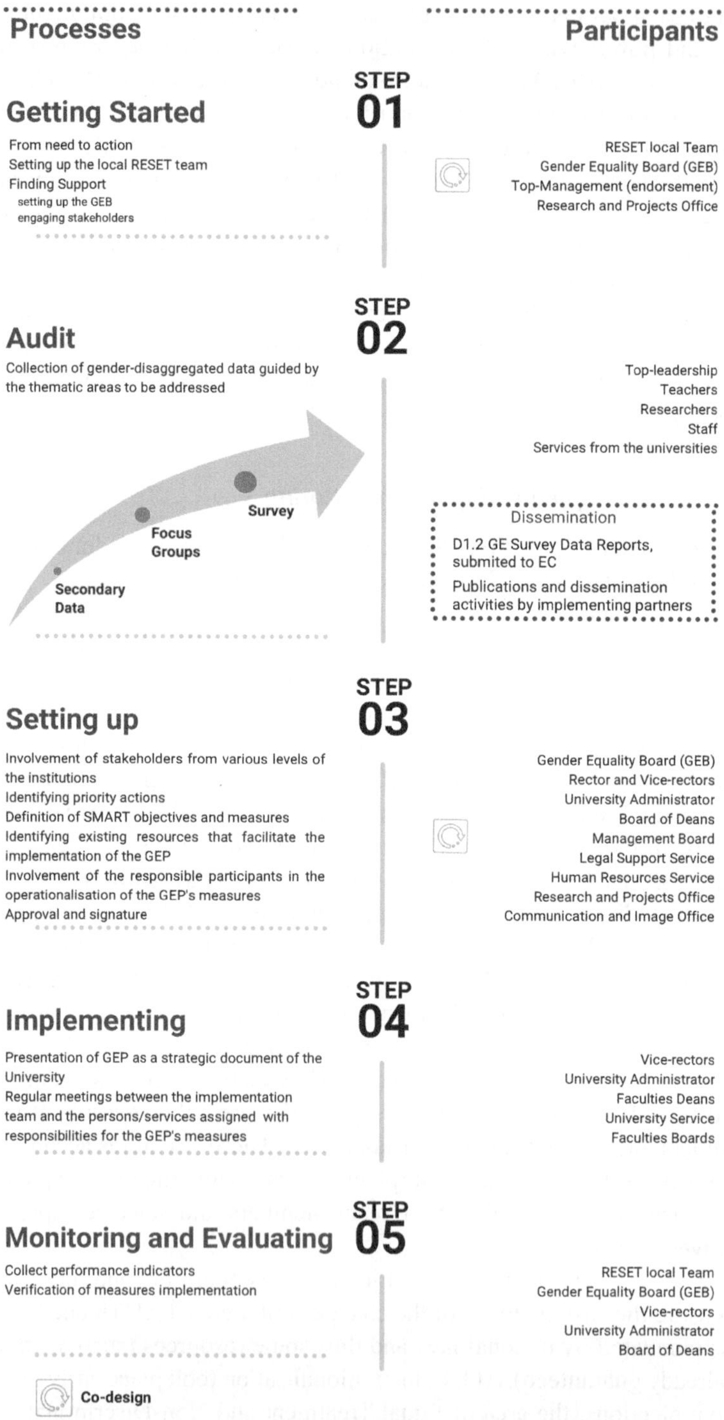


**FIGURE 6.1** GEPs development guided by GEAR tool and the co-design approach

under the Rectory's jurisdiction. At U.Porto, institutionalisation only occurred in the final stages of GEP implementation.

Moreover, to fully implement a mainstreaming strategy, a transformation of institutional and organisational cultures is key, and this is a process that requires changes in the policy process, its mechanisms, and the actors who take part (Lombardo & Mergaert, 2013). Indeed, while some partners have road maps and guiding documents on diversity, inclusion, or gender equality, besides GEPs for introducing GE (e.g., UBx), most workers remain unaware of them. A wide communication and dissemination strategy during the implementation step should be prioritised, for instance, by sending regular updates to the community on GEP actions, by developing campaigns, and by devoting sessions to GEP's state of implementation.

It is also important to be aware that not all measures of a first GEP were accomplished; if that was the case, either the GEP was designed with short ambition and no "big" measures were included or no risks were taken. Nonetheless, the goal of unfulfilled measures is to highlight the areas to change and to pinpoint the factors restricting change. This will aid in understanding the dynamics behind this mismatch and what needs to be done to achieve the measure even if it means the measure must be dropped. As such, one of the tools to increase the effectiveness of change is the **continuous collection of data and data analysis** that allows for a continuous cycle of diagnoses and monitoring. As this is a "back office" task, without major visibility *per se*, it is important to find ways to make it visible; the U.Porto team started publishing annual infographics with some key indicators informing the community and making data publicly available. This constant attention and data collection enabled not only minor adjustments and technical decisions but also a look forward on further steps needed allowing for clear intervention boundaries and responsibilities.

Implementing GE initiatives clearly requires context-specific approaches (Laoire et al., 2021), alongside structural elements such as data systems and dedicated resources.

### *Resetting GEPs: GE as a process*

Fully aware that the change process has been initiated with the GEP design, development, and implementation, it is always important to remember that new GEPs are to be devised. Indeed, the fact that an institution has a GEP in place does not, in itself, guarantee GE outcomes (O'Connor, 2020). Universities may engage in GEP actions for reasons other than a commitment to equality; for instance, because they have a legal duty to do so (Clavero & Galligan, 2021). Thus, while having a GEP in place is a positive element in seeking research funds, and in attracting talented staff and a greater pool of potential students, gender change and mainstreaming are lengthy processes.

Each new GEP is a step further to achieve greater mainstreaming, sustainability, and rooted GE. In the next section, we will take stock of the process developed in RESET's HEIs and reflect upon the barriers and challenges of gender mainstreaming.

## Reflect to action: how academic organisations can move equality further

After the experience of crafting, implementing, and evaluating the first GEPs in the four RESET partners, we summarise a set of reflections that can help develop further GEPs.

### *Resistance to GEPs: a multifaceted challenge*

One of the primary challenges to GEP implementation is resistance from various quarters within universities, at a time when gender issues generally tend to polarise public opinion in European societies. Resistance in this context can be defined as patterns of behaviour that people in power positions employ to actively or passively limit the implementation of gender equality initiatives (O'Connor & White, 2021). Some argue that GEPs are no longer necessary, claiming GE has been achieved or that existing policies suffice. Others advocate for equality beyond gender binaries, overlooking persistent disparities in academia, particularly in senior roles and specific fields (O'Connor, 2018).

Resistance also arises from concerns that focusing on gender excludes other important dimensions of inclusion, such as race or socio-economic status. LGBTQ+ groups, particularly trans and queer people, may critique GEPs as reinforcing binary gender norms, marginalising non-binary identities. While some push for universal equality, this undermines the need for targeted interventions to address systemic disadvantages faced by women and gender-marginalised groups. Another challenge is that resistance varies by university roles. Most academic roles are gendered, particularly those associated with power, which are predominantly masculine. However, resistance to gender equality initiatives is not exclusively gendered; for example, while some men may boycott or disengage from such efforts, women may also resist if they perceive their positions to be at risk due to these new measures. Therefore, it is important to adopt a multidimensional analysis of resistance, recognising that both men and women can be sources of resistance as well as allies in advancing GE. Furthermore, frontline staff, often witnessing microaggressions and discrimination, perceive greater inequities than senior leaders, who see improved gender balance at higher levels (Shober, 2014). For example, staff in student services often encounter micro-level discrimination, such as challenges faced by trans students in private spaces or in name use. Conversely, senior leadership might observe increased female representation in higher positions. This disconnection between leadership and frontline staff can also result in differing views on the necessity and relevance of GEPs.

### *Intersectionality: bridging the gap between policy and practice*

As mentioned, GEPs of the four universities were designed from an intersectional perspective – addressing not only gender but also factors such as age, ethnicity, sexual orientation, and socio-economic status. GEPs contained some measures that intersected gender with age, gender identity, LGBTQ+ rights, or professional group. However, implementing this approach has faced significant challenges. Institutions face structural and legal barriers that limit the collection of data on various forms of discrimination. Data disaggregated by race and ethnicity is non-existent, and gender data is often ambiguous, conflating sex and gender. Without such data, it becomes difficult to understand and address the complex layers of discrimination within academic institutions.

Moreover, identifying intersectional groups based on multiple identities could violate anonymity principles, posing both ethical and legal risks. Identifying precise intersection nodes could identify the people who fall within them and inadvertently expose individuals to further discrimination. It is, therefore, clear that the HEIs are not yet fully equipped to embrace an intersectional perspective. Despite these hurdles, the inclusion of intersectionality remains crucial for addressing the complex and overlapping forms of discrimination that students and staff may face.

While diversity policies target the entire community and aim to promote inclusion to all, these policies may clash with case-by-case and informal practices that, nevertheless, allow for a more personalised approach and address the specific needs of individuals – thus ensuring a truly intersectional perspective.

### *Diversity: expanding the scope while maintaining focus*

As mentioned, the concept of diversity presents both an opportunity and a challenge. The term diversity is, in fact, appealing to institutions because it allows them to overcome barriers relating to the concept of gender. Two discourses are often heard in academia: those who believe that the differences between men and women have been overcome or those who argue that there are no differences. Thus, the concept of diversity could extend the intervention of a GEP to all people, which better aligns with the concept of inclusion and, therefore, faces fewer barriers. Hence, it is undoubtedly an important project, but the risk of making such a broad plan is that it could dilute the gender inequalities that still exist in academia. Indeed, shifting an intersectional approach to a diversity one can obscure the overlapping forms of discrimination. While diversity policies promote visibility for groups defined by race, sexuality, or disability; targeting these groups in isolation undermines the intersectional approach.

At the consortium level, this concept was the subject of in-depth debate as reflected in the terminology "GE and Diversity" or simply "Diversity Equality".

In terms of inclusive language, there is a gradual acceptance of communication that includes both men and women, although language that is inclusive of non-binary identities remains less accepted. This reflects the ongoing tensions between traditional understandings of gender and more inclusive, non-binary approaches. This highlights the need for ongoing education and awareness-raising initiatives to promote a truly inclusive academic environment.

In sum, diversity thus allows to *Broaden the Conversation* fostering a more inclusive environment for all and/or *Dilute the Focus on Gender*, overshadowing the specific challenges faced by women, undermining the core objective of achieving GE.

### *Generation and socio-economic dimensions*

The academic environment mirrors broader societal changes, where different social classes and generations coexist, each with varying levels of engagement with gender issues. Older generations, having experienced more gender disparity, often see current progress as sufficient; while younger generations, influenced by social justice movements and digital activism, advocate for more radical, immediate changes. These younger generations, especially students, are more attuned to gender complexities and are supportive of LGBTQ+ rights and inclusive language. This generational divide reflects different understandings of equality, leading to varied expectations for achieving GE in academia.

Socio-economic status also influences perceptions of gender issues. Students and staff from lower socio-economic backgrounds may prioritise financial challenges over GE, viewing job stability and contract security as more pressing concerns. The intersection of class and gender adds complexity to GEP implementation, as not all members of the academic community are equally invested.

### *Sociopolitical climate*

The current sociopolitical climate offers both opportunities and challenges for GEPs. On the one hand, European funding regulations require universities to address gender imbalances, providing a financial incentive to GEP development. This external pressure creates a structural impetus for change, ensuring that GE is not only a matter of internal policy but also a requirement tied to institutional funding. In parallel, there is a growing awareness and advocacy for GE among younger generations and student bodies, who are more vocal and proactive in promoting inclusive practices, such as advocating for LGBTQ+ rights and for the use of inclusive language. On the other hand, resistance from right-wing ideologies and conservative backlash against social justice movements such as #MeToo complicates efforts to achieve GE. This highlights the need for HEIs to defend their

GE initiatives, balancing advocacy for progress with resistance from conservative elements.

## Conclusion

In reflecting on the experiences of crafting, implementing, and evaluating the first GEPs among the four RESET partners, it is evident that advancing GE in HEIs requires innovative and adaptive strategies. The complexities highlighted in this chapter underscore the need for nuanced approaches that address both persistent inequalities and emerging challenges. Resistance to GEPs remains a multifaceted challenge, rooted in differing perceptions of progress and concerns about inclusivity. While adopting an intersectional approach to GEPs holds promise for inclusivity, it also presents practical difficulties related to data collection, privacy, and institutional readiness. Similarly, while broadening the scope to embrace diversity can mitigate some resistance, it risks diluting the focus on specific gendered inequities that persist in academia.

To push gendered academic organisations further, it is crucial for HEIs to remain vigilant in defending their commitments to GE, while simultaneously navigating the broader dynamics of diversity and inclusion. Balancing targeted interventions for GE with a broader inclusive agenda will require sustained effort, reflection, and adaptation, ensuring that these plans evolve to meet the diverse needs of all members of the academic community.

## Note

1 https://eige.europa.eu/gender-mainstreaming/toolkits/gear/step-step-guide.

## References

Abels, Gabriele. (2011). Gender equality policy. In Hubert Heinelt & Michèle Knodt (Eds.), *Policies within the EU multi-level system*, 325–348. Nomos Verlagsgesellschaft mbH & Co. KG.

Bencivenga, Rita, Siri, Anna, Leone, Cinzia, & Taramasso, Angela C. (2021). Gender equality plans under Horizon Europe: From difficulties of interpretation to prospects. *AG About Gender – International Journal of Gender Studies, 10*(19), 378–397. https://doi.org/10.15167/2279-5057/AG2021.10.19.1316

Clavero, Sara, & Galligan, Yvonne. (2021). Delivering gender justice in academia through GE plans? Normative and practical challenges. *Gender, Work and Organization, 28*(3), 1115–1132. https://doi.org/10.1111/gwao.12658

European Commission. (2019). *She figures 2018*. Luxembourg: Publications Office of the European Union.

European Institute for Gender Equality. (2023). *GE index 2023: Towards a green transition in transport and energy*. Publications Office. https://data.europa.eu/doi/10.2839/64810

European Institute for Gender Equality (EIGE). (2024). Gender equality and gender mainstreaming. In *EIGE thesaurus*. https://eige.europa.eu/thesaurus/terms/1140

Forest, Maxime, & Lombardo, Emanuela. (2012). The Europeanization of gender equality policies: A discursive-sociological approach. In Emanuela Lombardo & Maxime Forest

(Eds.), *The Europeanization of GE policies. Gender and politics series*. London: Palgrave Macmillan. https://doi.org/10.1057/9780230355378_1

Hearn, Jeff, Strid, Sofia, Humbert, Anne L., Balkmar, Dag, & Delaunay, Marine. (2022). From gender regimes to violence regimes: Re-thinking the position of violence. *Social Politics, 29*(2), 682–705. https://doi.org/10.1093/sp/jxaa022

Iivari, Netta, Tervo, Erkki, Käsmä, Marjukka, & Heikkinen, Mervi. (2023). Participatory design meets gender equality at European higher education institutions. *Codesign, 19*(4), 304–326.

Laoire, Caitríona, Linehan, Carol, Archibong, Uduak, Picardi, Ilenia, & Udén, Maria. (2021). Context matters: Problematizing the policy-practice interface in the enactment of GE action plans in universities. *Gender, Work & Organization, 28*(2), 575–593. https://doi.org/10.1111/gwao.12594

Lätti, Johanna. (2017). Individualized sex equality in transforming Finnish academia. *European Educational Research Journal, 16*(2–3), 258–276. https://doi.org/10.1177/1474904116668184

Lombardo, Emanuela, Meier, Petra, & Verloo, Mieke. (2010). Discursive dynamics in gender equality politics: What about 'feminist taboos'? *European Journal of Women's Studies, 17*(2), 105–123. https://doi.org/10.1177/1350506809359562

Lombardo, Emanuela, & Mergaert, Lut. (2013). Gender mainstreaming and resistance to gender training: A framework for studying implementation. *NORA – Nordic Journal of Feminist and Gender Research, 21*(4), 296–311. https://doi.org/10.1080/08038740.2013.851115

O'Connor, Pat. (2018). Gender imbalance in senior positions in higher education: What is the problem? What can be done? *Policy Reviews in Higher Education, 3*(1), 28–50. https://doi.org/10.1080/23322969.2018.1552084

O'Connor, Pat. (2020). Why is it so difficult to reduce gender inequality in maledominated higher educational organizations? A feminist institutional perspective. *Interdisciplinary Science Reviews, 45*(2), 207–228. https://doi.org/10.1080/0308018 8.2020.1737903

O'Connor, Pat, & White, Kate. (2021). Power, legitimating discourses and institutional resistance to gender equality in higher education. In Pat O'Connor & Kate White (Eds.), *Gender, power and higher education in a globalised world. Palgrave studies in gender and education*. Cham: Palgrave Macmillan. https://doi.org/10.1007/978-3-030-69687-0_9

Paoletti, Marion, Quintin, Suzanne, Gray-Sadran, Jane, & Squarcioni, Laure. (2020). Sexual violence on campus objectification, awareness-raising and response. In Eileen Drew & Shiobán Canavan (Eds.), *The gender-sensitive university. A contradiction in terms?*, 67–78. London: Routledge.

Rixom, Jessica M., Jackson, Mark, & Rixom, Brett A. (2023). Mandating diversity on the board of directors: Do investors feel that gender quotas result in Tokenism or added value for firms? *Journal of Business Ethics, 182*, 679–697. https://doi.org/10.1007/s10551-021-05030-9

Roggeband, Conny, & Verloo, Mieke. (2006). Evaluating gender impact assessment in the Netherlands (1994–2004): A political process approach, *Policy & Politics, 34*(4), 615–632. https://doi.org/10.1332/030557306778553097

Rosa, Rodrigo. (2022). The trouble with 'work–life balance' in neoliberal academia: A systematic and critical review. *Journal of Gender Studies, 31*(1), 55–73. https://doi.org/10.1080/09589236.2021.1933926

Rosa, Rodrigo, & Clavero, Sara. (2022). Gender equality in higher education and research. *Journal of Gender Studies, 31*(1), 1–7. https://doi.org/10.1080/09589236.2022.2007446

Rosa, Rodrigo, Drew, Eileen, & Canavan, Siobhán. (2020). An overview of gender inequality in EU universities. In Eileen Drew & Sibhán Canavan (Eds.), *The gender-sensitive university. A contradiction in terms?*, 1–15. London: Routledge.

Shober, Dianne. (2014). Women in higher education in South Africa. In *Gender transformation in the academy* (Vol. 19, 315–332). Emerald Group Publishing Limited. https://doi.org/10.1108/S1529–212620140000019014

Vida, Bianka. (2020). Policy framing and resistance: Gender mainstreaming in Horizon 2020. European *Journal of Women's Studies*, *28*(1), 26–41. https://doi.org/10.1177/1350506820935495

Walby, Sylvia. (2005). Gender mainstreaming: Productive tensions in theory and practice. *Social Politics: International Studies in Gender, State & Society*, *12*(3), 321–343. https://doi.org/10.1093/sp/jxi018

# 7

# THE CHALLENGES OF DIVERSITY AND EQUALITY DATA IN ACADEMIC AND RESEARCH ORGANISATIONS

*Maria Michali, Areti Ampatzoglou, Eleni Karachaliou, Kalliopi Malerou, Stelios Karamanidis, Lamprini Argyropoulou, Dimitra Hadjipavlou-Litina, and Athena Vakali*

## Introduction

As academic and research organisations commit to fostering diverse and inclusive environments, the role of equality, diversity, and inclusion (EDI) data is considered essential to these efforts. Institutions are recognising the need to understand the composition and perspectives of their members more accurately to address systemic inequalities. In this context, data on gender, race, sexual orientation, socio-economic background, personal experiences, and other factors are considered crucial in supporting the implementation policies and progress measurement of diversity and inclusion initiatives. Studies have shown that understanding the specific challenges faced by individuals from intersecting minority identities enables organisations to develop more targeted and effective support measures, fostering a more inclusive environment overall (McKinsey & Company, 2021; World Economic Forum, 2023). However, organisations often encounter significant challenges in data collection, analysis, and application.

The collection of DEI data faces numerous methodological and ethical limitations. Variations in data collection practices across institutions can lead to inconsistencies, making it difficult to compare and assess progress effectively. Moreover, ethical considerations draw attention to the handling of sensitive information (Wolbring & Nguyen, 2023).

Beyond data collection, institutional barriers further obstruct the effective utilisation of DEI data. These barriers disrupt data exploitation to the end of designing and implementing interventions towards the elimination of discrimination and inequality. Resistance to change, particularly from established stakeholders who benefit from the status quo, as well as inadequate stakeholder engagement, can further hinder efforts to implement inclusive policies and strategies (Hodgins et al., 2022).

DOI: 10.4324/9781003564348-12

Current data-driven strategies are often limited in their effectiveness (Campbell & Helleloid, 2023). Many organisations rely greatly on quantitative data while overlooking qualitative insights into the experiences of stakeholders and marginalised groups. This imbalance can lead to ineffective initiatives that fail to address most inequality and exclusion issues.

To address these challenges, institutions recognise the need for innovative approaches to DEI data. Integrating qualitative and quantitative methods and leveraging advanced technologies, such as artificial intelligence (AI) and big data analytics, emerge as essential strategies for improving the collection and utilisation of diversity data, along with robust data governance frameworks (Luan et al., 2020). This chapter will explore the challenges and limitations of DEI data in academic settings. It will offer practical recommendations for overcoming these obstacles and enhancing the role of data in promoting inclusivity, aiming at providing valuable insights into advancing equality and equality within academic and research organisations.

## The role of diversity and equality data

### *Definition and types of DEI data*

DEI data encompass a range of information collected to understand and improve inclusivity within academic and research institutions. Collecting this type of data can provide the organisation with insights into the existing DEI situation and progress, and prominent challenges. DEI data are described as follows (Lunaria Solutions, n.d.):

- **Diversity data** reveals the demographic composition of an organisation's stakeholders, that is, professors, researchers, administrative employees, and students.
- **Equity data** reveals if these stakeholders have equal opportunities within the organisation, whether in employment, volunteer roles, services, products, or contracts.
- **Inclusion data** indicates if stakeholders feel included in their interactions with the organisation.

DEI data can be further, broadly categorised into demographic data, experiential data, and institutional metrics.

- **Demographic data** includes quantitative measures such as race, ethnicity, gender, age, disability status, sexual orientation, and socio-economic background. These metrics provide a snapshot of the composition of an institution's population, helping to identify under-represented groups and monitor changes over time (The EW Group). In the UK, data collection has expanded to include aspects such as faith, caring responsibilities, and disabilities (PwC, 2024).

- **Experiential data** captures qualitative insights into individuals' experiences within the institution. This can include surveys, interviews, and focus groups that explore perceptions of inclusivity, instances of discrimination or bias, and overall satisfaction with the institutional conditions. Such data is crucial for understanding the subtle realities of different groups and identifying areas for improvement.
- **Institutional data** pertains to policies, practices, and structures that impact EDI. This indicatively includes the analysis of hiring practices, promotion rates, pay equality, retention rates, and the presence of various DEI initiatives. Evaluating these metrics helps to identify systemic barriers and the effectiveness of policies in promoting an inclusive environment.

Collecting and analysing these data types require careful consideration of ethical standards and privacy concerns, ensuring the protection of individuals' identities and the integrity of the data. Data should be anonymised and analysed to determine if an institution is representative of the local community, identify which groups are in the majority or under-represented, and evaluate the impact of diversity initiatives (The EW Group). This is of major importance since, in recent years, organisations have evolved and intensified their approach to DEI due to societal changes and emerging regulatory mandates (PwC, 2024). This shift has led to a stronger emphasis on diversity data collection, enabling organisations to gain deeper insights (PwC, 2024). By comprehensively understanding and utilising these diverse data types, academic and research institutions can similarly develop targeted strategies to enhance DEI and encourage a more inclusive environment.

### *Importance of data in promoting inclusivity and equality*

Establishing effective objectives towards the promotion of DEI often demands a data-driven analysis to comprehend the socio-demographic profile of the organisation and identify any factors contributing to potential disparities in representation. Data analysis can similarly contribute to assessing how specific policies and action points may enhance equality in opportunities and, therefore, encourage diversity in the institution (PwC, 2024). In this context, data can assist organisations in identifying potential representation gaps and disparities, that is, areas where certain groups may be under-represented or experiencing unequal treatment. By recognising these gaps, institutions can proceed to targeted interventions to address inequalities (Helman et al., 2020). Institutional policymakers can also make informed decisions regarding policies and initiatives; evidence-based approaches ensure that efforts to promote inclusivity and equality are based on empirical data rather than assumptions (Harver, 2020; Helman et al., 2020).

Moreover, regular collection and analysis of DEI data allow organisations to monitor their progress over time and commit to change efforts. Ongoing progress assessments are essential for evaluating the effectiveness of DEI initiatives and making necessary adjustments to improve outcomes (Helman et al., 2020; PwC, 2024). The public disclosure of DEI data can further strengthen the organisational commitment to transparency and accountability. By sharing such information, institutions foster trust and credibility among stakeholders, openly demonstrating their dedication to advancing inclusivity and equality (PwC, 2024; Harver, 2020). Furthermore, detailed diversity data can reveal the specific needs of various demographic groups within the organisation. This enables the provision of customised support and resources that address the unique challenges faced by these groups, thus enhancing inclusivity (Ely & Thomas, 1996). Similarly, insights from diversity data can inform organisational strategies, including recruitment strategies for attracting more diverse candidates (Harver, 2020; Marshall et al., n.d.). Creating an inclusive environment based on data can also improve retention rates by ensuring that all employees feel valued and supported (Harver, 2020). Finally, detailed DEI data helps institutions comply with legal and regulatory requirements on equality and non-discrimination (EC, 2022; Equinet).

As for experiential data, these constitute a valuable tool for exploring the perspectives of the members of an organisation, acquiring detailed input for context-based policymaking, and ultimately enabling an equitable workplace culture (Homan et al., 2020). The qualitative nature of these data allows an in-depth examination of the institutional environment, including both formal and informal practices, and how these are experienced by the members. Data collected by specific qualitative methods, such as interviews, are considered especially reliable means for examining organisational environments (Cassell & Symon, 2012) and complex organisational realities (Eby et al., 2009), including DEI issues (see Peterson & Jordansson, 2022; Voorspoels & Bleijenbergh, 2019; Holck, 2016; Van den Brink & Stobbe, 2014). Such methods allow the inclusion of diverse voices (Eisenhardt & Graebner, 2007) considering that diverse individuals will be interviewed, for example, from different positions in the hierarchy. The inclusion of diverse perspectives considerably mitigates the risk of developing "one-size-fits-all" policies. It further mitigates researcher bias and any potential attempts on behalf of the participants to engage in impression management (Graebner & Eisenhardt, 2004). Impression management refers to an actor's attempts to create or sustain an image held for them by the target audience (Bozeman & Kacmar, 1997), usually a socially desirable one (Bolino et al., 2008).

### *Examples of data-driven initiatives in academic settings*

In academic settings, gathering and analysing DEI data is essential for shaping inclusive practices and policies. The RESET (Redesigning Equality and Scientific Excellence Together) project, funded by the EU under Horizon 2020, offers an example of how data can be used to guide the design and implementation of Gender Equality Plans (GEPs) across multiple institutions. This section outlines valuable data sources that can support similar initiatives, along with the value of such data in fostering gender equality and diversity.

*Key data sources for DEI initiatives*

- **Open data platforms**: Various publicly available datasets, such as those from **SheFigures** and **Eurostat**, provide gender equality indicators primarily for the EU context. These platforms offer comparative data on gender representation across different sectors and countries, making them useful for understanding broader trends in equality and diversity.
- **Institutional HR data**: Collecting gender and diversity data from HR departments within universities provides insights into the internal gender composition of academic and administrative staff. This type of data is crucial for identifying disparities and understanding the structural barriers to gender equality within institutions.
- **Surveys and focus groups**: Conducting surveys or focus groups within an institution allows for the collection of qualitative data that captures the lived experiences of staff. These can address issues of gender equality, inclusion, and workplace culture. Feedback from open-ended questions in these surveys can reveal more sophisticated challenges that quantitative data may not fully capture.

*The value of DEI data*

By leveraging these diverse data sources, institutions can develop a clearer understanding of the gender-related challenges they face. Data collection not only highlights areas for improvement but also supports evidence-based decision-making in the development of policies and practices aimed at improving gender equality and inclusion.

In addition, visualising institutional data through online platforms, such as the RESET project's Dashboard, helps to make DEI data accessible and actionable. Dashboards enable stakeholders to engage with real-time data, identify gaps, and monitor progress towards gender equality goals.

## Challenges in data collection and methodologies

Collecting DEI data in academic and research organisations is a crucial endeavour, yet there are numerous challenges. These challenges can undermine data accuracy, reliability, and ethical integrity (Gurcan, 2023; Mertens, 2023; Si et al., 2023), ultimately affecting the effectiveness of relevant initiatives. This section firstly explores frequently encountered obstacles in data collection and data methodologies, also experienced in the context of RESET, including issues of accuracy and reliability, ethical considerations, and technological and resource constraints.

Then, challenges experienced especially in the collection of qualitative experiential data are described.

### *Accuracy and reliability issues*

Accurate and reliable data are the cornerstones of any meaningful analysis. However, from the perspective of diversity and equality in academic settings, several factors hinder the collection of such data. These factors may include inconsistent data collection practices and the limited scope and granularity of the data collected.

One of the most significant challenges in data collection is the inconsistency in practices across different institutions or departments. This inconsistency can derive from a lack of standardised protocols for collecting diversity-related data. For example, some institutions may collect detailed demographic information, while others may only gather basic data or none. Additionally, the criteria used to classify this data can vary, leading to discrepancies that complicate efforts to compare or aggregate data across different settings. Inconsistent data collection practices can also result from differing levels of awareness and commitment to diversity issues. In some cases, data collection might be superficial, driven by the need to comply with reporting requirements. This can lead to incomplete data collection efforts, further diminishing the reliability of the data.

Moreover, data collection practices may differ due to varied technical capabilities and resources across institutions. Some organisations may lack the tools or expertise to collect and manage complex datasets effectively, leading to gaps or errors in the data. The result is an inconsistent data landscape that hinders comprehensive analysis and makes it difficult to draw meaningful conclusions or implement effective interventions.

Another challenge is the limited scope and granularity of the data that is often collected. Many data collection efforts focus on easily quantifiable aspects of diversity, such as gender and race, while neglecting more subtle dimensions, such as intersectionality. This limited scope can conceal important disparities and trends, preventing a full understanding of the diversity landscape.

Moreover, the data collected is often not granular enough to provide meaningful insights. For example, data might be aggregated in ways that de-emphasise differences within groups, such as not separating data by rank or role within the institution. This lack of granularity can lead to an oversimplified understanding of diversity issues and hinder efforts to address specific needs or challenges faced by particular subgroups.

### *Ethical considerations*

Ethical considerations are essential in the collection of diversity and equality data, given the sensitive nature of the information involved. Privacy and confidentiality concerns, and issues related to informed consent and data ownership, present significant challenges that must be carefully addressed.

Privacy and confidentiality are major ethical issues in the collection of DEI data. Individuals may be reluctant to provide personal information, particularly if they are unsure how it will be used. This concern is especially critical in environments where there is a lack of trust in how data will be managed or where there is a history of discrimination or bias. Maintaining confidentiality is also a significant challenge, particularly in settings where individuals may be easily identifiable, even in anonymised datasets.

Institutions must therefore implement effective data protection measures to ensure that the information collected is secure and that individuals' identities are protected. This is essential for building trust and encouraging participation in data collection efforts.

Informed consent is another ethical consideration in the collection of diversity data. Individuals must be fully informed about the purpose of the data collection, how their data will be used, and the measures to protect their privacy before they agree to participate.

### *Technological and resource constraints*

Technological and resource constraints often hinder the effective collection of DEI data. These constraints can limit the availability of data collection tools and the capacity of organisations to gather, manage, and analyse data effectively. The availability and accessibility of data collection tools vary significantly across institutions. Some may have access to sophisticated software and platforms for data collection and analysis, while others may rely on outdated or inadequate tools to handle complex diversity data. Moreover, some tools are not suitable for DEI data collection. For example, standard HR software might not include fields for capturing certain demographic information or may be incapable of handling the intersectional analysis required to understand diversity issues. These issues can lead to inconsistencies in data quality, limit the scope of the data collected, and hinder efforts to standardise data collection practices.

Finally, collecting and analysing diversity data requires both financial resources and skilled personnel. However, many academic and research organisations operate with limited budgets and may lack the necessary expertise or staff capacity to carry out comprehensive data collection efforts. Additionally, limited budgets may prioritise other areas of organisational operations, leaving diversity data collection activities underfunded. This can result in insufficient data collection processes, ultimately undermining data accuracy and reliability.

### *Subjectivity, bias, and generalisability*

Experiential data were also collected within RESET, to gain an informed view of the perspectives and experiences of the university population. Apart from the rich information gained by such qualitative data, some challenges and limitations were also encountered. These relate to the nature of qualitative data as such, and to the collection and analysis of qualitative data for DEI issues.

Qualitative data inherently exhibit some limitations about reliability. Reliability relates to concerns about whether several researchers analysing qualitative data can reach the same findings and conclusions (Silverman, 2007; Easterby-Smith et al., 2008). Reliability issues can be reinforced by various biases. Bias associated with qualitative data firstly encompasses researcher bias and subjectivity. All researchers bring their inherent subjectivity to the research process and are informed by previous experiences; thus, data collection, analysis, and interpretation can be affected accordingly. In qualitative methods, such as interviews and focus groups, researcher bias may arise from the interviewer's behaviour or questioning style (e.g., tone of voice or phrasing), inadvertently influencing how participants respond. Participant bias (interviewee bias) can occur when respondents withhold information, especially sensitive details, or modify their responses to present themselves in a favourable way, driven by social desirability concerns (Krumpal, 2013). The data triangulation strategy was adopted to mitigate these biases (Harvey, 2018). Qualitative insights were complemented by additional insights gained through large-scale surveys addressing a wider university sample.

A further challenge encompasses generalisability, and whether qualitative findings and insights can be replicated in other contexts (Lewis & Ritchie, 2003). Caution is needed, though, so that assumptions about generalisability do not fall under positivist notions since the aim of qualitative data is, in fact, to provide rich and context-based information about the questions addressed (Silverman, 1989). Therefore, generalisability may not be a major concern when the precise aim is to gain context-based insights in each university, as was the case in RESET.

### *Additional ethics concerns for DEI qualitative data*

Finally, qualitative, experiential data for DEI research are associated with additional ethics concerns. For instance, when conducting interviews, focus groups, or participant observation for DEI research, there are risks related to psychological pressure or re-traumatisation experienced by participants, or even reputational risks (Faas et al., 2022). In addition, prominent concerns relate to confidentiality and privacy (Gui, 2023). Participants in DEI research most often share sensitive information; thus, they may hesitate to participate in case they are identifiable. As for participant diversity, respect for cultural sensitivity needs to be ensured (Brayda & Boyce, 2014), for instance, when interacting with individuals who have different cultural values and norms. Power dynamics between researchers and participants similarly evoke concerns (Bengtsson & Fynbo, 2018), especially when people from minority groups, whose voices are usually marginalised, participate in qualitative studies.

To overcome this concern, a potential strategy is to keep any sensitive insights and information – potentially making a participant identifiable – private. Moreover, when conducting focus groups, the aim should be to create a welcoming and inclusive environment to inspire openness and confidentiality.

## Institutional barriers to inclusive practices

Apart from challenges inherent in data and respective methods for analysis, additional constraints stem from the institutional context. The extensive collection of EDI-related demographic and institutional data (including historical ones) requires a robust institutional infrastructure for data collection, particularly in large organisations (Fisher et al., 2024). Thus, considerable time and resources must be devoted to ensure the existence of such infrastructure (Carrilio et al., 2004). Concurrently, organisations may lack a robust strategy background to provide guiding directions for such data-driven initiatives (PwC, 2024). Such operational obstacles can be encountered particularly in newly adopted gender policies and interventions, such as GEPs, where organisations need to invest in unprecedented and extensive data collection efforts.

Data collection efforts can be further hindered by resistance (Carrilio et al., 2004). Organisational members may be reluctant, or refuse to participate in interviews, focus groups, and other data-collection activities (Gui, 2023). This form of covert resistance may indicate opposition to change or DEI progress. Alternatively, it may reflect a general fear of involvement and being exposed, stemming from concerns about how data will be used or perceived. This situation was also encountered within RESET, especially when focus groups were conducted. Overall, both cases potentially result in data collection efforts that are confined to a small, familiar cohort of participants; consequently, the pool of data is either insufficient or reflects partial experiences and perspectives. For instance, as Lombardo and Bustelo (2022) argue, reporting protocols for harassment incidents that were not consistently used by the members of a Spanish university evoked a lack of data on harassment cases.

Finally, organisational over-reliance on data, mostly statistical data and quantifiable metrics, may prevent employees from reporting difficult situations, such as harassment, that cannot be captured by data (Campbell & Helleloid, 2023). Similarly, organisational services responsible for employee affairs, such as HR departments, may require data-driven evidence to intervene, overall leaving some DEI problems unaddressed (Salmon, 2022).

Overall, these barriers underscore the complexity of implementing systematic data collection initiatives within institutional settings, especially when navigating issues of stakeholder involvement. Genuine support of top management is essential in these cases in order to prevent adverse effects on effective data exploitation (Dorsey, 2002).

## Limitations of current data-driven strategies

Based on the challenges mentioned earlier, various limitations can emerge for organisational data-driven strategies. This section additionally incorporates insights from the RESET project, highlighting how challenges in DEI data collection and utilisation can significantly constrain the design and implementation of effective, inclusive strategies. Inadequate collection, analysis, and interpretation of DEI data, either quantitative or qualitative, can lead to poor and ineffective decision-making and strategy-making. For instance, inconsistent collection methods and the absence of robust data quality standards may result in data that varies significantly in accuracy and reliability, making it difficult to draw meaningful conclusions. Similarly, outdated, or incorrect quantitative data, can lead to flawed insights.

In addition to these issues, the absence of a robust data management infrastructure, such as a centralised system for data storage and organisation, makes it challenging to access and analyse data efficiently. This situation not only hinders the ability to derive meaningful insights but also poses security risks, as sensitive information may not be adequately protected. Such vulnerabilities highlight the need for secure, centralised, and well-structured data management systems that facilitate ethical and effective data usage.

Over-reliance on quantitative data at the expense of qualitative data can also lead to neglecting the experiences of organisational members (Campbell & Helleloid, 2023); qualitative data may occasionally be more difficult to collect but are vital for ensuring effective interventions.

The limitations of data-driven strategies that build on qualitative data still need to be considered, though. For instance, DEI qualitative data builds on individuals' perspectives and standpoints; these may reflect partial knowledge and a hegemonic interpretation of sensitive DEI issues, depending on individuals' background, status, and power. Thus, organisational interventions may selectively draw upon certain narratives that align with hegemonic ideologies, such as whiteness and heteronormativity, ultimately privileging specific groups (Harris, 2017) and not benefiting equally all organisational members, particularly those whose experiences do not align with dominant narratives.

Finally, limited analytical capabilities often hinder organisations' ability to extract valuable insights from the data. Without robust analytical tools and methods, organisations may fail to uncover key patterns and correlations within data that are crucial for developing effective strategies. Moreover, the reliance on basic or outdated analytical methods can result in a narrow understanding of stakeholder experiences and organisational dynamics. To overcome these barriers, organisations must invest in more sophisticated data analytics capabilities that can handle both quantitative and qualitative data effectively, ensuring a more inclusive and accurate representation of the issues at hand.

## Innovative approaches to data utilisation

### *Integrating qualitative and quantitative data*

Overall, both quantitative and qualitative DEI data provide valuable insights and contribute to promising data-driven strategies. Mixed-methods approaches, which utilise the strengths of both data types, can further provide highly beneficial insights and enhance evidence-based decision-making (Creswell & Plano Clark, 2017).

Mixed-methods approaches are particularly flexible, allowing either quantitative or qualitative data to take precedence while the study progresses, depending on the research objectives. Equal prioritisation of both data types is similarly feasible (Plano Clark, 2017). This iterative and flexible design is further enhanced by the possibility of data triangulation, where multiple data sources are used to mitigate bias and enhance valid conclusions (Fusch et al., 2018). Overall, moving beyond disputes about the benefits of qualitative as opposed to quantitative data or vice versa can enable an in-depth examination of DEI issues, including intersectional implications (Ben, 2018). In-depth discussions and feedback on DEI issues are similarly enhanced (Kolodziejczyk, 2015) since quantitative data reveal patterns, trends, and statistical relationships, while qualitative data provide nuanced insights into the lived experiences, perceptions, and attitudes of individuals. This was also verified in RESET, where the adopted mixed-methods approach enabled the nuanced comprehension of the organisational environment and associated DEI challenges. Consequently, it provided a valuable basis for the GEPs' development and implementation.

It is finally noted that, gradually, several studies move beyond the traditional qualitative/quantitative dichotomy and acknowledge the contribution of mixed-methods approaches in both academic settings and beyond. For instance, Nelson and Zippel (2021) combined qualitative and computational methods to explore how the concept of implicit bias is translated into practice within a US-funded university programme. They collected qualitative data through document and text analysis of proposal documents, interviews, and participant observations, which were then subjected to computational analysis, including semantic network analysis and word frequency analysis. This approach explored how implicit bias manifests at macro, meso, and micro levels: in other words, across the funding landscape, within the specific programme, and among the programme's participants. In addition, Abrahamsson (2014) conducted a longitudinal study to examine the gender-related impact of organisational change projects in Swedish industrial organisations. Interviews and participant observations were utilised in the initial study, while a follow-up study also included an analysis of company statistics. Authors were thus able to examine organisations' efforts across time to adopt modern organisational models and address gender segregation.

### *Leveraging technology and big data*

Technology and big data further provide promising avenues for DEI in organisational and academic settings. DEI integration is a factor already accounted for in research, with gradually increasing attention. Embedding inclusive research design and methodologies not only increases awareness but can also enhance the quality and accuracy of research, and provide valuable input for DEI initiatives. Several apps, websites, and datasets already support this effort, including:

- AI Fairness 360 by IBM: The AI Fairness 360 is an extensible open-source toolkit which can help the user examine, report, and mitigate discrimination and bias in machine learning models throughout the AI application life cycle. It includes a comprehensive set of metrics for datasets and models to test for biases, explanations for these metrics, and algorithms to reduce bias in datasets and models.
- Gender Decoder: This tool was inspired by a research paper written by Gaucher et al. (2011). The site provides a quick way to check whether a job advert has a kind of subtle linguistic gender-coding.
- Fairlearn (Microsoft Research): Fairlearn was started in 2018 by Miro Dudik from Microsoft Research as a Python package to accompany the research paper of Agarwal et al. (2018). The tool has been further developed and is now an open-source, community-driven project assisting data scientists in enhancing the fairness of AI systems.
- Trinka AI [Inclusive Language Check]: Trinka AI's Inclusive Language Check is an advanced tool that helps users identify and eliminate potential biases in writing. It analyses the content for six key categories of bias: gender bias, disability bias, racial bias, religious bias, nationality bias, and age bias.
- Gender Data Portal by World Bank|Kaggle|Gender Statistics Database by EIGE|Data.world: These platforms host a diverse range of datasets, including those focused on DEI across various sectors. Users can use these resources to support their efforts in creating more inclusive research environments.

The earlier tools can aid in ensuring that DEI data are more valuable and accurate. For instance, additional information and datasets can be retrieved, enriching the data pool. Moreover, bias inherent in data and AI systems used along DEI initiatives can be identified and mitigated.

### *Best practices for data governance and strategy*

To ensure the effective collection and utilisation of DEI data in organisational contexts, a carefully planned data strategy and governance framework are essential. Also leveraging experiences from the RESET project, this section highlights best practices designed to address common limitations and overcome potential organisational barriers. First, compliance with national and EU-level regulatory frameworks is needed to uphold ethics, integrity, and principles of confidentiality and/or anonymity. Standardised data collection practices are equally important to ensure the consistency and comparability of data over time. Continuous monitoring and evaluation of collected DEI data is similarly needed to acquire constructive feedback on DEI practices and implement any necessary refinements. In this way, iterative and genuine progress towards organisational goals can be achieved (Advance HE, 2022).

In addition, building trust with organisational stakeholders is vital in order to ensure that they feel comfortable sharing their data. Teams responsible for the data collection and analysis can foster stakeholder trust by providing clear, detailed information on the purposes of data collection and analysis, and being available to provide clarifications and guidance as needed. Support from top management and commitment to DEI progress is equally important, facilitating access to and retrieval of specific data types, like historical ones. It can further enable institutional legitimacy for DEI efforts, which similarly facilitates data collection efforts.

Finally, inclusive data collection can provide reliable and inclusive insights, and mitigate the exclusion of marginalised groups. This can be achieved by distributing widely and to various university stakeholder groups' anonymous surveys and invitations for focus group discussions. These include various hierarchies and age groups, including academic, research, teaching and administrative staff, and students.

## Conclusion

This chapter has highlighted the critical role of DEI data in promoting an inclusive academic and research environment. The findings underline the importance of collecting and utilising DEI data to address inequalities in academic institutions and ensure context-based interventions. However, academic institutions face challenges, such as inconsistent data collection practices, ethical concerns, and institutional resistance, that hinder the effective use of data. Furthermore, the overreliance on quantitative data, at the expense of qualitative insights, can result in inadequate or ineffective interventions.

The future of DEI data in academic institutions requires compliance with national and EU-level regulatory frameworks in terms of ethics, integrity, and confidentiality. Furthermore, innovative methods that integrate quantitative and qualitative approaches, as well as the use of advanced technologies, such as AI and big data analytics, provide significant opportunities for improving the accuracy and effectiveness of data. This perspective can help institutions gain a more comprehensive understanding of the experiences of marginalised groups, thereby enabling the application of more effective policies and strategies.

It is essential for stakeholders – including university leadership, policymakers, and staff – to genuinely prioritise DEI efforts. This includes thoroughly designed data strategy and governance, addressing internal resistance and lack of engagement, and embracing innovative methodologies for data collection and analysis. By prioritising these initiatives, academic and research institutions can take significant steps to create more fair and inclusive environments for all members of their communities.

## References

Abrahamsson, Lena. (2014). Gender and the modern organization, ten years after. *Nordic Journal of Working Life Studies*, *4*(4), 109–136. https://doi.org/10.19154/njwls.v4i4.4710

Advance HE. (2022). *Guidance on the collection of diversity monitoring data*. https://s3.eu-west−2.amazonaws.com/assets.creode.advancehe-document-manager/documents/advance-he/Advance%20HE%20Guidance%20on%20the%20collection%20of%20diversity%20monitoring%20data%20final_1713881453.pdf

Agarwal, Alekh, Beygelzimer, Alina, Dudík, Miroslav, Langford, John, & Wallach, Hanna. (2018). A reductions approach to fair classification. *Proceedings of the 35th International Conference on Machine Learning*, *80*, 60–69.

Ben, Esther Ruiz. (2018). Intersectionality in the practice of mixed methods gender research. *Journal of Research in Gender Studies*, *8*(1), 73–88.

Bengtsson, Tea Torbenfeldt, & Fynbo, Lars. (2018). Analysing the significance of silence in qualitative interviewing: Questioning and shifting power relations. *Qualitative Research*, *18*(1), 19–35. https://doi.org/10.1177/1468794117694220

Bolino, Mark C., Kacmar, K. Michele, Turnley, William H., & Gilstrap, J. Bruce. (2008). A multi-level review of impression management motives and behaviors. *Journal of management*, *34*(6), 1080–1109. https://doi.org/10.1177/0149206308324325

Bozeman, Dennis P., & Kacmar, K. Michele. (1997). A cybernetic model of impression management processes in organizations. *Organizational Behavior and Human Decision Processes*, *69*(1), 9–30. https://doi.org/10.1006/obhd.1996.2669

Brayda, Winsome Chunnu, & Boyce, Travis D. (2014). So you really want to interview me?: Navigating "sensitive" qualitative research interviewing. *International Journal of Qualitative Methods*, *13*(1), 318–334. https://doi.org/10.1177/160940691401300115

Campbell, Katherine, & Helleloid, Duane. (2023). Ann Hopkins and PwC: Critically analyzing culture, leadership, policies, and reporting for diversity and inclusion. *Issues in Accounting Education*, *38*(1), 185–199. https://doi.org/10.2308/ISSUES-2021–115

Carrilio, Terry E., Packard, Thomas, & Clapp, John D. (2004). Nothing in – nothing out: Barriers to the use of performance data in social service programs. *Administration in Social Work*, *27*(4), 61–75. https://doi.org/10.1300/J147v27n04_05

Cassell, Cathy, & Symon, Gillian. (2012). *Qualitative organizational research: Core methods and current challenges*. Thousand Oaks: SAGE Publications.

Creswell, John W., & Plano Clark, Vicki L. (2017). *Designing and conducting mixed methods research*. Thousand Oaks: SAGE publications.

Data.world. (n.d.). *The data catalog built for your AI future*. https://data.world/

Dorsey, David W. (2002). Information technology. In Jerry Hedge & Elaine D. Pulkalos (Eds.), *Implementing organizational interventions*, 110–132. San Francisco: Jossey-Bass.

Easterby-Smith, Mark, Thorpe, Richard, & Jackson, Paul. (2008). *Management research* (3rd ed.). London: SAGE.

Eby, Lillian T., Hurst, Carrie S., & Butts, Marcus M. (2009). The redheaded stepchild in organizational and social science research. In Charles E. Lance & Robert J. Vandenberg (Eds.), *Statistical and methodological myths and urban legends: Doctrine, verity and fable in the organizational and social sciences*, 219–246. New York: Routledge.

Eisenhardt, Kathleen M., & Graebner, Melissa E. (2007). Theory building from cases: Opportunities and challenges. *Academy of Management Journal*, *50*(1), 25–32. https://doi.org/10.5465/amj.2007.24160888

Ely, Robin J., & Thomas, David A. (1996). Making differences matter: A new paradigm for managing diversity. *Harvard Business Review*, *74*(5), 79–90.

Equinet Europe. (n.d.). *EU legislative framework*. https://equineteurope.org/equality-in-europe/eu-legislative-framework/

European Commission, European Education and Culture Executive Agency. (2022). *Towards equality and inclusion in higher education in Europe*. Luxembourg: Publications Office of the European Union.

The EW Group. (n.d.). *How to collect diversity data.* https://theewgroup.com/blog/collect-diversity-data/

Faas, Simone, Makhija, Simrin, Bryan, Elizabeth, Go, Ara, & Malapit, Hazel. (2022). *Considering gender in research: An ethics and standards toolkit.* Nairobi: CGIAR GENDER Impact Platform. https://hdl.handle.net/10568/121973

Fairlearn Community. (n.d.). *Fairlearn.* https://fairlearn.org/

Fisher, Oliver J., Fearnshaw, Debra, Watson, Nicholas J., Green, Peter, Charnley, Fiona, McFarlane, Duncan, & Sharples, Sarah. (2024). Promoting equality, diversity and inclusion in research and funding: Reflections from a digital manufacturing research network. *Research Integrity and Peer Review, 9*(1), 5. https://doi.org/10.1186/s41073-024-00144-w

Fusch, Patricia, Fusch, Gene E., & Ness, Lawrence R. (2018). Denzin's paradigm shift: Revisiting triangulation in qualitative research. *Journal of Sustainable Social Change, 10*(1), 19–32. https://doi.org/10.5590/JOSC.2018.10.1.02

Gaucher, Danielle, Friesen, Justin, & Kay, Aaron C. (2011). Evidence that gendered wording in job advertisements exists and sustains gender inequality. *Journal of Personality and Social Psychology, 101*(1), 109–128. https://doi.org/10.1037/a0022530

Graebner, Melissa E., & Eisenhardt, Kathleen M. (2004). The seller's side of the story: Acquisition as courtship and governance as syndicate in entrepreneurial firms. *Administrative Science Quarterly, 49*(3), 366–403. https://doi.org/10.2307/4131440

Gui, Tianhan. (2023). Me too? Yes, me too! Sexual harassment of female expatriates in China's overseas enterprises. *Violence Against Women, 29*(11), 2288–2310. https://doi.org/10.1177/10778012221125497

Gurcan, Fatih. (2023). What issues are data scientists talking about? Identification of current data science issues using semantic content analysis of Q&A communities. *PeerJ Computer Science, 9*, e1361. https://doi.org/10.7717/peerj-cs.1361

Harris, Kate Lockwood. (2017). Re-situating organizational knowledge: Violence, intersectionality and the privilege of partial perspective. *Human Relations, 70*(3), 263–285. https://doi.org/10.1177/0018726716654745

Harver. (2020). *The state of diversity recruiting in 2020.* https://harver.com/wp-content/uploads/2020/03/White-paper_-The-state-of-diversity-recruiting-v8.pdf

Harvey, Steven A. (2018). Observe before you leap: Why observation provides critical insights for formative research and intervention design that you'll never get from focus groups, interviews, or KAP surveys. *Global Health: Science and Practice, 6*, 299–316. https://doi.org/10.9745/GHSP-D-17–00328

Helman, Alex, Bear, Ashely, & Colwell, Rita. (2020). *Promising practices for addressing the underrepresentation of women in science, engineering, and medicine: Opening doors.* Washington, DC: The National Academies Press.

Hodgins, Margaret, O'Connor, Pat, & Buckley, Lucy-Ann. (2022). Institutional change and organisational resistance to gender equality in higher education: An Irish case study. *Administrative Sciences, 12*(2), 59. https://doi.org/10.3390/admsci12020059

Holck, Lotte. (2016). Putting diversity to work: An empirical analysis of how change efforts targeting organizational inequality failed. *Equality, Diversity and Inclusion: An International Journal, 35*(4), 296–307. https://doi.org/10.1108/EDI-12-2015-0107

Homan, Astrid C., Gündemir, Seval, Buengeler, Claudia, & van Kleef, Gerben A. (2020). Leading diversity: Towards a theory of functional leadership in diverse teams. *Journal of Applied Psychology, 105*(10), 1101–1124. http://dx.doi.org/10.1037/apl0000482

IBM Research. (n.d.). *AI Fairness 360.* https://aif360.res.ibm.com/

Kaggle. (n.d.). *Level up with the largest AI & ML community.* https://www.kaggle.com/

Kolodziejczyk, Iwona. (2015). Mixed methods for study of gender issues in access, application, and attitudes toward ICT in higher education institutions in Papua New Guinea. *Sage Open, 5*(2), 1–15. https://doi.org/10.1177/2158244015581017

Krumpal, Ivar. (2013). Determinants of social desirability bias in sensitive surveys: A literature review. *Quality & Quantity*, *47*(4), 2025–2047. https://doi.org/10.1007/s11135-011-9640-9

Lewis, Jane, & Ritchie, Jane. (2003). Generalising from qualitative research. In Jane Ritchie & Jane Lewis (Eds.), *Qualitative research practice*, 109–113. London: SAGE Publications.

Lombardo, Emanuela, & Bustelo, Maria. (2022). Sexual and sexist harassment in Spanish universities: Policy implementation and resistances against gender equality measures. *Journal of Gender Studies*, *31*(1), 8–22. https://doi.org/10.1080/09589236.2021.1924643

Luan, Hui, Geczy, Peter, Lai, Hollis, Gobert, Janice, Yang, Stephen J. H., Ogata, Hiroaki, Baltes, Jacky, Guerra, Rodrigo, Ling, Pi, & Tsai, Chin-Chung. (2020). Challenges and future directions of big data and artificial intelligence in education. *Frontiers in Psychology*, *11*, 1–11. https://doi.org/10.3389/fpsyg.2020.580820

Lunaria Solutions. (n.d.). *Diversity data collection*. https://lunariasolutions.com/blog-post/deidata/

Marshall, Ernest W., Hirsch, Wendy, & Eaton, Mei Kim. (n.d.). Improving diversity with data. *Society for Human Resource Management*. https://www.shrm.org/executive-network/insights/people-strategy/improving-diversity-data

McKinsey & Company. (2021). Women in the workplace 2021. *LeanIn.Org and McKinsey & Company*. https://www.mckinsey.com/featured-insights/diversity-and-inclusion/women-in-the-workplace

Mertens, Donna M. (2023). *Research and evaluation in education and psychology: Integrating diversity with quantitative, qualitative, and mixed methods*. Thousand Oaks: SAGE publications.

Nelson, Laura K., & Zippel, Kathrin. (2021). From theory to practice and back: How the concept of implicit bias was implemented in academe, and what this means for gender theories of organizational change. *Gender & Society*, *35*(3), 330–357. https://doi.org/10.1177/08912432211000335

Peterson, Helen, & Jordansson, Birgitta. (2022). Gender mainstreaming in Swedish academia: Translating policy into practice. *Journal of Gender Studies*, *31*(1), 87–100. https://doi.org/10.1080/09589236.2021.2004539

Plano Clark, Vicki L. (2017). Mixed methods research. *The Journal of Positive Psychology*, *12*(3), 305–306. https://doi.org/10.1080/17439760.2016.1262619

PwC. (2024). *Diversity data guide 2024: Additional guidance on the collection and use of data to enhance workforce EDI*. https://www.pwc.co.uk

Salmon, Udeni. (2022). "It's wicked hard to fight covert racism": The case of microaggressions in science research organizations. *Gender, Work & Organization*, *31*(3), 727–748. https://doi.org/10.1111/gwao.12934

Si, Li, He, Yi, & Liu, Li. (2023). Topics and changing characteristics of knowledge organization research in the 21st century: A content analysis. *Journal of Documentation*, *79*(2), 487–508. https://doi.org/10.1108/JD-05-2022-0101

Silverman, David. (1989). Telling convincing stories: A plea for cautious positivism in case-studies. In Barry Glassner & Jonathan D. Moreno (Eds.), *The qualitative-quantitative distinction in the social sciences*, 57–77. Dordrecht, Netherlands: Springer.

Silverman, David. (2007). *A very short, Fairly interesting and reasonably cheap book about qualitative research*. London: SAGE.

TrinkaAI. (n.d.) *Inclusive language check*. https://www.trinka.ai/features/inclusive-language-check

Van den Brink, Marieke, & Stobbe, Lineke. (2014). The support paradox: Overcoming dilemmas in gender equality programs. *Scandinavian Journal of Management*, *30*(2), 163–174. https://doi.org/10.1016/j.scaman.2013.07.001

Voorspoels, Jolien, & Bleijenbergh, Inge. (2019). Implementing gender quotas in academia: A practice lens. *Equality, Diversity and Inclusion: An International Journal*, *38*(4), 447–461. https://doi.org/10.1108/EDI-12-2017-0281
Wolbring, Gregor, & Nguyen, Annie. (2023). Equity/equality, diversity and inclusion, and other DEI phrases and DEI policy frameworks: A scoping review. *Trends in Higher Education*, *2*(1), 168–237. https://doi.org/10.3390/higheredu2010011
World Bank Group. (n.d.). *Gender data portal*. https://genderdata.worldbank.org/en/home
World Economic Forum. (2023). *Diversity, equity, and inclusion: An overview of best practices for organizations*. https://www.weforum.org/reports

# 8

# GENDER-INCLUSIVE COMMUNICATION IN ACADEMIC ORGANISATIONS

## A lever for change

*Viktoria Niebel*

### Introduction

Gender-inclusive communication practices sometimes attract attention as "trigger points" (Mau et al., 2023) in European societies and political discourse (Mayer, 2021; Kováts, 2021). Two recent examples showing the political and institutional impact of these controversies are the measures that came into effect in the German states of Bavaria and Hesse. In April 2024, the Free State of Bavaria banned the use of gender-inclusive language in Bavarian authorities, schools, and universities.[1] This particularly affects the use of the gender asterisk, colon, or medial dot, which are the most common signs to use gender-inclusive language in German. The Bavarian state government justified the ban by stating that language must be clear and understandable and argued that ideologically influenced forms of language, such as gendering, could have an exclusionary effect. The ban applies to all official written communications, including official documents, internal communications, and educational materials. Violations may result in disciplinary consequences. Also, in April 2024, the bans in Hesse restricted the use of gender-inclusive language in schools.[2] Students are not allowed to use special characters like the gender asterisk, colon, or underscore in exams; these are marked as errors and graded accordingly. Both states base their measures on the recommendations of the German Spelling Council and argue the necessity of clear and comprehensible language.

The regulations enacted in both states not only illustrate the political dimension of gender-inclusive and diversity-oriented communication but also highlight the controversy that accompanies the use or restriction of gender-sensitive language (Simon, 2022; Waldendorf, 2024). In contrast to the initial examples in 2 of 16 German Federal Estates, in the UN's guidelines on gender-inclusive language, practices of gender-inclusive communication are explicitly highlighted as a tool to

DOI: 10.4324/9781003564348-13

promote gender equality, as language has a crucial role in shaping society and culture (UN, n.d.). Furthermore, the European Parliament (2008) adopted multilingual guidelines that foster gender-neutral and gender-inclusive language (GIL) in 2008.

The rejection of scientific and political commitment to gender equality measures is a global phenomenon (Mayer, 2021; Kováts, 2021). However, it is not reserved for a specific political agenda or affiliation. Populist, conservative, religious, and far-right movements, discourse strategies, and lines of argumentation often share the rejection of gender-inclusive practices of communication as a response against the so-called "gender ideology" (Kuhar & Paternotte, 2017; Paternotte & Kuhar, 2018). According to these views, "gender ideology" and its manifestations (as in gender-sensitive practices of communication) are considered artificial, a devaluation of and a threat to traditional, heteronormative structured family forms and gender roles, and even a threat to languages in their supposed integrity (Vergoossen et al., 2020). At the same time, this discourse against gender equality and against the recognition of gender and sexual diversity is often mixed with the rejection of other social and cultural forms of diversity, making European societies heterogeneous, versatile, and complex constellations. As such, they cannot be reduced to a supposed "true cultural identity" addressed by the metaphor of cultural "roots" (Bettini, 2016; Mayer, 2021).

Against this background, the commitment to gender-inclusive and diversity-oriented communication takes on a political dimension and responsibility, as the examples from Bavaria and Hesse show. Hesitations towards inclusive communication practices, therefore, can arise from many reasons and different sources and can vary in their extent. They are also expressed by people who have had little exposure to its concepts and rationale, or who are confused by the ambiguities of the multiple possibilities of gender-equitable language. These people are often part of our institutional and organisational environment, work with us, and are involved in the practices and actions of communication in Higher Education Institutions (HEIs). It is therefore of great importance to raise awareness and share information on the aims and possibilities that come along with GIL, and to debunk myths around gender-inclusive communication practices (Leal et al., 2024). Thus, the goal of advancing social participation regardless of gender is expressed in measures that promote gender-inclusive communication. It is in response to this goal that the Horizon 2020 funded project *Redesigning Equality and Scientific Excellence Together* (RESET) focused on improving institutional communication practices and developed the *Toolbox on Gender-neutral, and gender-inclusive institutional communication* (Niebel, 2022). In this chapter, issues of gender-inclusive institutional communication are discussed in connection to the notion of togetherness in HEIs. Against the background of feminist institutionalism, the example of the toolbox will be outlined, and central, sociopsychological aspects of gender-inclusive communication will be highlighted and discussed. This chapter focuses on gender-sensitive and diversity-oriented language as an aspect of organisational culture, as fundamental understandings of togetherness can be found in these approaches: they share the concern to perform representation and address in written and visual form. In this way, they counter exclusion and invisibility in words and images.

## Feminist institutionalism and the notion of togetherness in academic institutions

Universities are complex social spaces where individuals from diverse economic, cultural, and social backgrounds come together and interact within relationships that, depending on various configurations and roles, are often shaped by hierarchical structures. Gender is one socio-structural and demographic factor among others that shapes the diversity of social situations and individual and collective experiences of inequality at academic institutions and society as a whole.

At the same time, universities, with their campus communities and self-proclaimed, as well as societally demanded, mission, are envisioned as institutions that promote cultural democracy, citizenship, and civic participation through the free and open collaboration of all individuals, grounded in their unique personalities and contributions (Procentese et al., 2019; Ostrander, 2004). This vision embodies the idea that "becoming and being an academic citizen" fosters individuals' growth into more engaged and informed political citizens (Englund, 2002, p. 286). Therefore, the university community facilitates the ability to actively participate in both national and international cultural dialogues (Englund, 2002; Procentese et al., 2019).

From a perspective rooted in feminist institutionalism, universities can also be understood as organisations shaped by structures, processes, and power dynamics (Niebel & Heikinnen, 2025). These institutions are not considered gender-neutral; rather, they are deeply gendered in their operations and hierarchies that have been fundamentally structured around male norms and experiences (Acker, 1990, 2006). Acker's perspective challenges traditional organisational theories that treat institutions as abstract entities, detached from the gendered and racialised bodies that inhabit them. Following this approach, the gendering of organisations – such as universities – can be examined across various spheres of institutional life. Thus, institutional norms can be seen as deeply intertwined with organisational culture (Clavero & Galligan, 2020), revealing the pervasive influence of gender on the functioning and structure of academic institutions. Measures of RESET that target to change academic environments are based on these understandings of universities as gendered organisations. *Togetherness* in this context means developing policies and practices that actively prevent discrimination and foster equal opportunities for everyone within the institution. RESET's toolbox for gender-neutral and diversity-oriented communication is one example of this understanding of institutional action.

It is the notion of *togetherness* that inspires the collaborative and culture-centric approach of RESET, and therefore also centres the role of communication as a lever for gender equality and diversity in the project. The notion of togetherness emphasises collaboration, empathy, and shared experiences as keys to fostering community (Hannikainen & van Oers, 1999). Whether in classrooms, work groups, or social gatherings, togetherness is built on mutual support, cooperation, and a recognition of shared goals. It is not dependent on a particular activity but arises from a collective participation that binds individuals into a unified whole. This broader understanding of togetherness underscores its role in promoting inclusion, belonging, and cohesion within educational and social contexts, making it a crucial element for fostering collaborative and inclusive environments in higher education and beyond.

Togetherness in higher education and academic institutions means creating a supportive and equitable environment where diverse perspectives are not only acknowledged but actively valued, allowing all individuals to flourish both academically and socially. In the following, key elements of the social and cultural psychological perspective on the relevance of fostering inclusive communication practices in academic institutions will be discussed.

## Social and cultural psychology of communication

Gender-sensitive and diversity-oriented communication not only responds to an ideal of equal participation, representation, and recognition, as outlined in the previous sections but is also based on social and psychological knowledge about the effects and functioning of communication. It promotes inclusion over exclusion. Communication can also be understood as communicative action and can take the form of language as well as visual dimensions, and a variety of other forms of expression (e.g., body language and posture, facial expressions, gestures, touch, and clothing; Broszinsky-Schwabe, 2016, p. 32; Reichertz, 2020). Communication also assumes the function of conveying meaning in social interactions, aims at understanding, and is processual, and relational (Burkhart, 2003). Furthermore, communication always takes place in complexly structured contexts. Various factors interact with these, such as the setting and environment of the communication, its place, its occasion, and the time at which communication takes place (Broszinsky-Schwabe, 2016, p. 33). Accordingly, communication is always entangled in discursive contexts of meaning.

Especially in our highly variable settings at universities, communication takes on a variety of forms, so written announcements, oral speeches, and (audio-)visual representations all need to be reflected upon, and institutional practice should ensure that it becomes as inclusive as possible. For example, we can think of job advertisements in which the wording does not simply characterise the job profile and present the competencies of ideal applicants, but itself forms a counter-gendered image of the competence requirements. This often has unintended effects on the recruitment process even before the application process has been completed (Horvath & Sczesny, 2015).

The aspect of conveying meaning should be explored in greater depth at this point: from a psycholinguistic perspective, in particular, it has been shown that gender-inclusive language has a significant impact on the equality of all genders in our thinking and perception, and ultimately on our actions (Horvath & Sczesny, 2015; Stout & Dasgupta, 2011). Therefore, the promotion of gender-inclusive communication is an important tool to foster gender equality.

Following Hall (2013), we understand language as a construct that goes beyond the oral or written processing and dissemination of information and the constitution of meaning. Language thus also takes on its expression in visual materials or bodily practices. Communication is a social practice, and as such, it is also culturally shaped and normatively loaded. Thus, it is never neutral and its normativity can vary (Cavanaugh, 2020). This implies that language functions as a reflection of personal and social ideas and structures. The use of language not only determines our thinking and our perception but also has a decisive influence on our actions. It "is also and always infused with and caught up in the political economic, national, (post)colonial, and political circumstances that shape its use and its role as an object of study, political manipulation, and cultural value" (Cavanaugh, 2020, p. 51). Thus, language also creates meaning and reality. At the same time, it constantly changes.

### *Stereotypes and gender bias*

The intentional use of language in the sense of gender-sensitive and gender-neutral communication is both a field of action for gender equality measures, as well as a political point of debate in European societies, as has been exemplified at the very beginning of this chapter. As such, it aims to challenge stereotypes, address people of different genders equally, and make diversity visible, which is inherent to our universities and our societies.

*Stereotypes* and *prejudices* are common to all kinds of social interactions, and also influence these in HEIs. They arise when individuals are viewed primarily as members of social groups rather than being recognised for their individual characteristics (Straub & Niebel, 2021). They influence how individuals perceive others as well as themselves based on presumed group affiliations, such as gender, age, ethnicity, sexual orientation, or religious background (Hannover & Wolter, 2017). Often, multiple social categories intersect, amplifying the effects of these stereotypes. Also, cultural norms and values play a significant role in shaping the content of gender stereotypes (Cuddy et al., 2015).

While prejudices reflect attitudinal judgements towards others, stereotypes function as cognitive frameworks that categorise both individuals and groups (Straub & Niebel, 2021). These mental structures are part of how we process social information and can influence how we perceive others and how we view ourselves in relation to gender (Hannover & Wolter, 2017). For example, a female student entering a male-dominated field, such as engineering, might have internalised social stereotypes that suggest men are, in a way, more naturally suited for technical tasks. As a result, she may begin to question her own abilities and competencies, and have the impression that she needs to prove her competencies over and over again. This internalisation of gendered attributes can cause her to feel as though she doesn't fully belong in the field, even if her skills are objectively strong, and they can also have an influence on her self-perception and performance even if they are quite the opposite of her attitudes and values.

*Biases*, which are closely linked to stereotypes, represent another critical aspect of social cognition. They are defined as the conscious or unconscious interpretation or representation of a phenomenon that consistently favours one perspective over another (Chandler & Munday, 2020). These biases often manifest implicitly, shaping attitudes and behaviours towards certain individuals or groups, which ties them to prejudicial thinking. Biases can "be manifested in behavior as discrimination; or in representation, the evaluative loading of description or depiction so that connotative meanings become definitive characteristics" (Chandler & Munday, 2020). In the context of gender, biases take on particular significance. Gender bias refers not only to explicit prejudices about gender roles but also to unconscious processes that shape communication practices and representations. Specifically, gender bias can manifest in how communication or representation adheres to the male norm, especially within patriarchal societies, where the dominant worldview is often presented as neutral or universal, even though it is shaped by those in power (Chandler & Munday, 2020). This creates a subtle, often invisible, reinforcement of gendered norms, affecting interactions and representations in higher education settings.

### *Gender-inclusive and gender-exclusive communication*

Language is not merely a neutral medium of communication; it carries implicit biases that can perpetuate stereotypes and exclusions. As discussed, biases manifest in communication through patterns that favour certain groups, particularly men, which can also be observed in the everyday use of language. This is why *Gender-sensitive language* emphasises making all genders – women, men, and non-binary individuals – visible by directly addressing them and actively avoiding stereotypical or exclusionary terms. In contrast, *gender-neutral language* seeks to avoid gendered references altogether, using terms that are intended to be as inclusive as possible by making gender invisible in the discourse (EIGE, 2019).

These strategies fall under the broader umbrella of *gender-inclusive language* (GIL), an approach for ensuring equal representation in communication. Gender-sensitive language explicitly incorporates all gender identities, challenging traditional norms and reducing the risk of reinforcing stereotypes through exclusionary wording. However, gender-neutral language, while attempting to offer a more egalitarian approach, can inadvertently reinforce gender bias by making gender invisible. This approach risks perpetuating traditional assumptions by not actively addressing all genders, as people might default to masculine representations in their interpretations.

This is evident in the common use of the *generic masculine*, a form of gender-discriminatory language (Waldendorf, 2024). It involves the use of masculine nouns and grammatical structures as the default for all people, implicitly including women and non-binary individuals without explicitly acknowledging them. This asymmetry, where masculine forms are considered "neutral" and female forms are used only when specifically referring to women, reflects a deep-rooted bias in language (Sczesny et al., 2016).

For example, consider reading a report on a scientific conference that details the number of participants, the focus of the content, and other key aspects. In languages that use the generic masculine, it remains unclear whether women were among the lecturers and speakers. The text only makes men visible, leaving the presence of women and non-binary individuals ambiguous (Niebel, 2022).

As a result, the ordinary and habitual use of the generic masculine contradicts the principles of equal participation and gender inclusivity. When masculine forms are used generically, it is often unclear whether the statement truly includes all genders or only refers to men, as the information remains implicit. This ambiguity not only reinforces traditional gender hierarchies but also obscures the visibility of non-masculine identities in everyday communication (Stout & Dasgupta, 2011). Therefore, language that follows the generic masculine model perpetuates the very gender bias that gender-inclusive language seeks to dismantle.

### *Visual communication*

Visual communication plays a crucial role in shaping institutional narratives and perceptions, particularly in the context of diversity and gender equality. As Gunnarsson (2019) points out,

> Images – or the absence of images – affect how we see each other, and ourselves. When some people are invisible in the media, when their voices and stories are never heard, there is a void left to be filled with prejudice and stereotypes.

In alignment with this, RESET's toolbox underscores the importance of visual communication in fostering inclusive environments. This involves not only ensuring the visibility of people of colour, individuals with disabilities, those with migration backgrounds, LGBTIQ communities, and elderly individuals but also doing so in a way that avoids reinforcing stereotypes or tokenism. The goal of visual representation is to reflect the diversity of our institutions and societies (Maier, 2018).

However, while visibility is a crucial aspect of promoting inclusion, it also presents inherent challenges. Visibility does not automatically equate to empowerment (Hall, 2013; Mirza, 2018). For example, visibility can lead to exposure, which can be both appreciative and vulnerable at the same time. In academic settings, increased visibility for marginalised groups, especially women of colour, can often result in their being objectified as "mute visible objects" (Mirza, 2018), without granting them any real increase in power or voice. Moreover, the desire for recognisability can inadvertently lead to the formation of new stereotypes, even when attempting to break free from negative ones (Hall, 2013). This dynamic highlights the complexity of visual representation, as how minorities are depicted can often fall into the trap of tokenism.

*Tokenism* points to the practice of promoting individuals from marginalised groups into high-profile roles to create the illusion of equal opportunity, without addressing the underlying structures of inequality (Heery & Noon, 2008). When applied to visual communication, tokenism risks instrumentalising individuals from disadvantaged groups as symbols of diversity without fostering genuine change. The depiction of diversity in institutional communication, therefore, carries the danger of being reduced to a performative act, where marginalised individuals are showcased to signal inclusivity, while systemic biases and exclusions remain unaddressed.

## RESET toolbox for gender-neutral and diversity-oriented communication

The *Toolbox for Gender-Neutral, Diversity-Oriented Institutional Communication* developed within the RESET project aimed to promote sustainable cultural change by fostering a gender- and diversity-friendly environment within HEIs. The toolbox focused on establishing gender equality and diversity through an intersectional lens and therefore addressed multiple axes of discrimination, including but not limited to gender, and targets students, researchers, university management, and administrative staff. The toolbox was built upon the insights from sister projects, such as the SUPERA project's Guidelines for Gender-Sensitive Communication in Research and Academia (2021), and EIGE's toolkit on gender-sensitive communication (2019). The methodological approach of the toolbox was grounded in a co-designing process (Iivari et al., 2023) with all RESET partner universities being involved by discussing and co-designing new strategies for fostering inclusive communication based on their practices and experiences. The creation of the toolbox was driven by an iterative process that began with a thorough evaluation of the current state of gender-inclusive communication at each institution. Because this process can be of relevance to other stakeholders interested in developing such language and communication tools, the toolbox encompasses these evaluation questionnaires and interview scripts as tools (Niebel, 2022). Indeed, by assessing the existing, institutional practices, concerns, and aspirations regarding gender-inclusive language and visual communication, it is possible to gain insight into whether gender-inclusive and diversity-sensitive communication was actively being discussed and implemented, or if it remained a marginal topic in the institutional communication. Importantly, these initial interviews marked the beginning of a continuous process that not only shaped the development of the toolbox but also provided a foundation for its future evolution, ensuring that the strategies for inclusive communication remain responsive to institutional needs and challenges.

When preparing for this toolbox, we aimed at complementing a conceptual framework (part A) with practical tools (part B). Thus, this toolbox first provides reasoning that explains the need for gender-sensitive, gender-neutral, and diversity-oriented communication, detailing RESET's intersectional understanding of diversity within HEIs. It also explores the socio-psychological and sociolinguistic aspects of communication related to gender and diversity, as well as the opportunities and challenges posed by visual communication, as discussed in Section 3. A set of practical tools, designed through institutional queries, interviews, and discussions, serve as a foundation for creating more inclusive communication practices, addressing both gender and broader diversity factors.

The toolbox targets various actors and institutions within universities. It primarily addresses those who are aware of the potential for change in current communication practices and are in a position to develop appropriate measures. This includes gender equality and diversity officers, as well as staff in communication departments, organisational development, and human resources, who can promote gender equality and inclusivity in their administrative roles. These actors play a key role in establishing a gender-equitable and diversity-sensitive communication culture and ensure that the importance of these aspects is also reflected in the transfer of scientific knowledge to society. Thus, the toolbox provides both a theoretical framework and practical approaches to help identify the need for change in institutional communication and to inspire the further development of such processes. A particular focus is placed on visual language as an integral part of institutional communication, as visual representations significantly influence how gender roles and diversity are perceived and conveyed in academic contexts. In addition, the toolbox addresses key aspects of gender-inclusive communication in academia, particularly concerning scientific writing and the organisation of conferences. This objective underscores that the promotion of gender equality and diversity is not solely an institutional process but also needs to permeate scientific practices themselves to foster a long-term inclusive and equitable academic culture.

## Discussion and conclusion

In higher education settings, where clear, equitable communication is essential, adopting gender-inclusive communication practices becomes particularly important. By moving away from exclusive practices, educational institutions can foster an environment that values and reflects the diversity of all individuals, thereby promoting a more inclusive and egalitarian space for learning, teaching, and researching. This issue touches on broader societal dynamics, particularly regarding the intersection of power structures and representation. As Vertovec (2012) articulates, visual representation is entangled with issues of recognition, redistribution, and organisation, all of which are critical in understanding how institutional power manifests. While the intention to depict diversity may arise from a genuine desire for inclusion, it is essential to remain aware of the risk of perpetuating exclusionary practices, such as appropriation or the display of superficial "happy faces" in the attempt to perform inclusive communication (Ahmed, 2009). True progress in institutional communication requires confronting the mechanisms of racism, inaccessibility, and other forms of exclusion, while actively working to counteract stereotypes and biases.

In this context, visual communication must move beyond surface-level inclusion and engage with the deeper processes that shape representation. It should aim to create spaces where diversity is displayed and meaningfully integrated into the institutional fabric, ensuring that marginalised voices are not only seen but also heard and empowered.

In reflecting on the journey towards implementing gender-inclusive and diversity-sensitive communication within academic institutions, it is important to recognise both the sociolinguistic and socio-psychological dimensions that shape this process, which is why these dimensions have been highlighted in this text. As explored in previous sections, verbal and visual language plays a critical role in how gender and diversity are represented and perceived in institutional settings. Coming back to the notion of togetherness, two significant factors have been identified in RESET's analysis leading to the communication toolbox: resistance and safety (Niebel, 2022).

Resistance to change is a well-documented phenomenon in organisational behaviour, particularly in relation to gender equality initiatives. Resistance often arises as an attempt to maintain the status quo when new gender-inclusive communication practices are introduced (Benshop & Verloo, 2011). In the context of higher education, this resistance can manifest in various forms, for example, from a lack of prioritisation of gender-sensitive communication to outright opposition, such as aversion to neologisms or the belief that gender equality has already been achieved. These attitudes reflect a deeper reluctance to challenge entrenched norms, such as the use of the generic masculine as a "neutral" form, for example, in German or in Latin languages (Leal et al., 2024). In HEIs, these resistances occur at organisational, operational, and individual levels, and are often intertwined (FESTA, 2016). Acknowledging these barriers is essential for developing effective strategies to counteract them. That is why RESET's toolbox offers practical responses to common forms of resistance (Niebel, 2022).

However, it is not enough to merely counter resistance. To foster gender-inclusive communication, the notion of *safety* must also be prioritised. As Dembouski (2018), Mirza (2018), and Lester (2018) point out, safety is deeply connected to diversity, particularly in academic environments where individuals experience the institution differently based on their gender, race, socio-economic status, and other intersectional factors. Safety in this context is not merely the absence of harm but the creation of spaces where all individuals feel seen, heard, and respected.

Designing communication practices and environments that explicitly address the challenges and rules of communication, to be inclusive, is the first step in promoting this sense of safety. And it goes further: creating safe spaces for articulation, where individuals can express their needs without fear of judgement or marginalisation, is equally important. This includes making support systems – whether financial aid, disability services, or resources for first-generation students – both accessible and clearly communicated. It also involves ensuring that institutional policies and rules, particularly those related to studying, researching, and working, are transparent and inclusive. Teachers and staff must be trained to recognise student's diverse backgrounds and experiences and to address factors that may influence classroom interactions or assessments beyond academic content.

The need for safety in institutional communication extends beyond formal policies – it also demands a shift in perspective: While much of the discourse around gender-inclusive communication focuses on institutional responsibilities, it is essential to consider the lived experiences of individuals within these institutions. For many, particularly those from marginalised or vulnerable groups, the need for change is felt most acutely in everyday interactions, in being overlooked or silenced. This is where communication plays a vital role in shaping an environment where everyone, regardless of gender or background, feels valued and supported.

As institutional, inclusive communication practices continue to evolve, the ongoing dialogue will remain essential. This work is inherently a process – one that requires productive, critical exchange among university members, with particular attention to those whose experiences of exclusion or invisibility demand change. RESET's *Toolbox for Gender-Neutral, Diversity-Oriented Institutional Communication* integrated these reflections on resistance and safety into the tools provided, and thereby aims to support universities in creating environments where gender and diversity are not only respected but actively nurtured. This is not just a matter of policy; it is about reshaping the cultural and communicative practices that define our academic institutions and making them places of genuine inclusion and belonging.

## Notes

1 https://www.sueddeutsche.de/bayern/gendern-sternchen-verbot-sprache-bayern-1.6468805 (13.09.2024).
2 https://www.zeit.de/gesellschaft/2024–03/gendersprache-verbot-hessen-landesverwaltung (13.09.2024).

## References

Acker, Joan. (1990). Hierarchies, jobs, and bodies: A theory of gendered organizations. *Gender & Society, 4*, 139–158.

Acker, Joan. (2006). Inequality regimes. *Gender & Society, 20*(4), 441–464. https://doi.org/10.1177/0891243206289499

Ahmed, Sara. (2009). Embodying diversity: Problems and paradoxes for black feminists. *Race Ethnicity and Education, 12*, 41–52.

Benshop, Yvonne, & Verloo, Mieke. (2011). Gender change, organizational change, and gender equality strategies. In E. Jeanes, D. Knights, & P. Yancey-Martin (Eds.), *Handbook of gender, work and organization*, 277–290. John Wiley.

Bettini, Maurizio. (2016). *Radici: Tradizione, identità, memoria*. Bologna: Il Mulino.

Broszinsky-Schwabe, Edith. (2016). *Interkulturelle Kommunikation. Missverständnisse und Verständigung*. Springer.

Burkhart, Roland. (2003). Kommunikation als soziale Interaktion. In J. Bolten & C. Erhardt (Eds.), *Interkulturelle Kommunikation. Texte und Übungen zum interkulturellen Handeln*, 17–42. Verlag Wissenschaft & Praxis.

Cavanaugh, Jillian R. (2020). Language ideology revisited. *International Journal of the Sociology of Language, 263*, 51–57.

Chandler, Danie, & Munday, Robert. (2020). *A dictionary of media and communication* (3rd ed.). Oxford University Press.

Clavero, Sara, & Galligan, Yvonne. (2020). Analysing gender and institutional change in academia: Evaluating the utility of feminist institutionalist approaches. *Journal of Higher Education Policy and Management, 42*(3), 1–17. https://doi.org/10.1080/1360080X.2020.1733736

Cuddy, Amy J., Wolf, Elizabeth B., Glick, Peter, Crotty, Susan, Chong, Jihye, & Norton, Michael. (2015). Men as cultural ideals: Cultural values moderate gender stereotype content. *Journal of Personality and Social Psychology, 109*(4), 622–635.

Dembouski, Lisa M. (2018). Honest and uncomfortable: A loving look at my exclusive campus. In S. A. Shelton, J. Ewing Flynn, & T. J. Grosland (Eds.), *Feminism and intersectionality in academia. Women's narratives and experiences in higher education*, 103–116. Palgrave Macmillan.

EIGE. (2019). *Toolkit on gender-sensitive communication. A resource for policymakers, legislators, media and anyone else with an interest in making their communication more inclusive*. European Institute for Gender Equality. https://eige.europa.eu/sites/default/files/20193925_mh0119609enn_pdf.pdf

Englund, T. (2002). Higher education, democracy and citizenship – the democratic potential of the university? *Studies in Philosophy and Education, 21*(4–5), 281–287. https://doi.org/10.1023/A:1019840006193

FESTA. (2016). *Handbook on resistance to gender equality in academia*. https://www.festa-europa.eu/sites/festa-europa.eu/files/FESTA%20D7.1%20Handbook%20on%20Resistance%20to%20Gender%20Equality%20in%20Academia.pdf

Gunnarsson, Tomas. (2019). *Images that change the world. A guide to equal communication*. Swedish Institute.

Hall, Stuart. (2013). The spectacle of the 'other'. In S. Hall, J. Evans, & S. Nixon (Eds.), *Representation*, 215–287. SAGE.

Hannikainen, Maritta, & van Oers, Bert. (1999). *Signs and problems of togetherness in a community of learners*. https://eric.ed.gov/?id=ED436241

Hannover, Bettina, & Wolter, Wolter. (2017). Geschlechtsstereotype: Wie sie entstehen und sich auswirken. In B. Kortendiek, B. Riegraf, & K. Sabisch (Eds.), *Handbuch interdisziplinäre Geschlechterforschung*. Springer.

Heery, Edmund, & Noon, Mike. (2008). *A dictionary of human resource management* (2nd rev. ed.). Oxford University Press.

Horvath, Lisa K., & Sczesny, Sabine. (2015). Reducing women's lack of fit with leadership positions? Effects of the wording of job advertisements. *European Journal of Work and Organizational Psychology, 25*(2), 316–328.

Iivari, Netta, Tervo, Erkki, Käsmä, Marjukka, & Heikkinen, Mervi. (2023). Participatory design meets gender equality at European higher education institutions. *Codesign – International Journal of Cocreation in Design and the Arts*. https://doi.org/10.1080/15710882.2023.2215742

Kováts, Eszter. (2021). Anti-gender politics in East-Central Europe: Right-wing defiance to West-Eurocentrism. *Gender, 1*, 76–90.
Kuhar, Roman, & Paternotte, David. (2017). *Anti-gender campaigns in Europe: Mobilizing against equality*. Rowman & Littlefield International.
Leal, Daniela, Peixoto Freitas, Jorge, Magalhães, Sara Isabel, & Matias, Marisa. (2024). Quem inclusivo fala, sempre acerta: Uma discussão sobre linguagem inclusiva. *Ex Æquo, 49*, 19–34. https://doi.org/10.22355/exaequo.2024.49.03
Lester, Aryah. (2018). Living with three strikes: Being a transwoman of color in education. In S. A. Shelton, J. Ewing Flynn, & T. J. Grosland (Eds.), *Feminism and intersectionality in academia. Women's narratives and experiences in higher education*, 145–152. Palgrave Macmillan.
Maier, Tanja. (2018). Von der Repräsentationskritik zur Sichtbarkeitspolitik. In R. Drüeke, E. Klaus, M. Thiele, & J. E. Goldmann (Eds.), *Kommunikationswissenschaftliche gender studies*, 77–90. Transcript.
Mau, Steffen, Lux, Thomas, & Westheuser, Linus. (2023). Triggerpunkte. Konsens und Konflikt in der Gegenwartsgesellschaft. Suhrkamp.
Mayer, Stefanie. (2021). Anti-gender-Diskurse – vom, gesunden Menschenverstand 'zur,Politik mit der Angst'. In S. A. Strube, R. Perintfalvi, R. Hemet, M. Metze, & C. Sahbaz (Eds.), *Anti-genderismus in Europa: Allianzen von Rechtspopulismus und religiösem Fundamentalismus. Mobilisierung – Vernetzung – Transformation*, 35–50. Transcript. https://doi.org/10.14361/9783839453155–003
Mirza, Heidi S. (2018). Decolonizing higher education: Black feminism and the intersectionality of race and gender. *Journal of Feminist Scholarship, 7*, 1–12. https://digitalcommons.uri.edu/jfs/vol7/iss7/3
Niebel, Viktoria. (2022). *Toolbox for gender-neutral, diversity-oriented institutional communication (2.0, update 2023)*. Zenodo. https://doi.org/10.5281/zenodo.8095962
Niebel, Viktoria & Heikinnen, Mervi. (2025). *Intersectional approaches to anti-sexism and anti-racism in German universities: A case study, ZDfm – Zeitschrift für Diversitätsforschung und -management*, (1)2025, 9–21. https://doi.org/10.3224/zdfm.v10i1.02
Ostrander, S. A. (2004). Democracy, civic participation, and the university: A comparative study of civic engagement on five campuses. *Nonprofit and Voluntary Sector Quarterly, 33*(1), 74–93. https://doi.org/10.1177/0899764003260588
Paternotte, David, & Kuhar, Roman. (2018). Disentangling and locating the 'global right': Anti-gender campaigns in Europe. *Politics and Governance, 6*(3), 6–19. https://doi.org/10.17645/pag.v6i3.1557
Procentese, Fortuna, Gatti, Flora, & Falanga, Annarita. (2019). Sense of responsible togetherness, sense of community and participation: Looking at the relationships in a university campus. *Human Affairs, 29*(2), 247–263. https://doi.org/10.1515/humaff-2019–0020
Reichertz, Jo. (2020). Einleitung. In J. Reichert (Ed.), *Grenzen der Kommunikation – Kommunikation an den Grenzen*, 11–27. Velbrück.
Sczesny, Sabine, Formanowicz, Magda, & Moser, Franziska. (2016). Can gender-fair language reduce gender stereotyping and discrimination? *Frontiers in Psychology, 7*, 25. https://doi.org/10.3389/fpsyg.2016.00025
Simon, Horst J. (2022). Sprache macht Emotionen. Geschlechtergerechtigkeit und Sprachwandel aus Sicht der historischen Soziolinguistik. *Aus Politik und Zeitgeschichte, 72*, 16–22.
Stout, Jane G., & Dasgupta, Nilanjana. (2011). When he doesn't mean you: Gender-exclusive language as ostracism. *Personality & Social Psychology Bulletin, 37*(6), 757–769. https://doi.org/10.1177/0146167211406434
Straub, Jürgen, & Niebel, Viktoria. (2021). *Kulturen verstehen, kompetent handeln*. Psychosozial-Verlag.

SUPERA. (2021). *Tailor-made guides for gender-sensitive communication in research and academia*. https://www.superaproject.eu/wp-content/uploads/2021/07/D8.2-TAILOR-MADE-guides-for-gender-sensitive-communication-in-research-and-academia-v.1.1-.pdf

United Nations. (n.d.). *Guidelines for gender-inclusive language in English*. https://www.un.org/en/gender-inclusive-language/guidelines.shtml

Vergoossen, Hellen P., Renström, Emma A., Lindqvist, Anna, & Gustafsson Sendén, Marie. (2020). Four dimensions of criticism against gender-fair language. *Sex Roles, 83*, 328–337. https://doi.org/10.1007/s11199-019-01108-x

Vertovec, Steven. (2012). 'Diversity' and the social imaginary. *European Journal of Sociology, 53*(3), 287–312. https://doi.org/10.1017/S000397561200015X

Waldendorf, Anica. (2024). Words of change: The increase of gender-inclusive language in German media. *European Sociological Review, 40*(2), 357–374. https://doi.org/10.1093/esr/jcad044

# PART III

# Bringing about cultural change

# INTRODUCTION TO PART III

## Inclusive excellence: bringing about cultural change through training

In this part of the book, consisting of four chapters, we will focus on the importance of training and capacity-building in bringing about institutional change and enhancing employees' skills and knowledge with regard to equality and diversity. Part III also closely tackles the crucial role of various stakeholders and CoPs (Communities of Practice) in HEIs in co-designing training activities according to the needs of particular academic groups and in disseminating knowledge and skills gained through training. Particular attention is given to the key role of trainers and their know-how in gender equality and diversity training, as well as to the timely need to introduce GIA tools (Gender Impact Assessment) in knowledge production.

In the first chapter, Aleksandra Różalska and Agata Rudnicka elaborate on the roles, responsibilities, and engagement of various academic stakeholders in the co-designing of the training programme and of the efforts to make it effective in bringing about sustainable change in institutional culture. Deriving from stakeholder theory by R. Edward Freeman and methodologically from Bloom's Taxonomy of Goals and Kolb's Cycle, the chapter offers an inclusive proposal of gender education for different stakeholders, taking into consideration their different needs and expectations, and using various teaching forms and methods based on the example of the training toolbox developed within the RESET project.

The second chapter by Mateusz Hauk and Paulina Szymańska draws from various theories that inform DEI (Diversity, Equality, and Inclusivity) training today. It emphasises the importance of including in the training process not only practical knowledge on such issues as discrimination, harassment, bias, and aggression but also individual emotional state of employees, effective communication skills,

DOI: 10.4324/9781003564348-15

building successful professional relationships, and, finally, the capability to recognise and react to inappropriate behaviours. In addition, the focus is on the role of trainers in the awareness- and knowledge-raising process and in building competencies and skills necessary to bring about institutional change. In her chapter on Gender Impact Assessment (GIA), Mervi Heikkinen discusses the importance of including gender and intersectionality in knowledge production, both in research and teaching conducted in HEIs. She argues that scholars should not omit these dimensions while planning their research, as they are fundamental aspects in any society, which influence all spheres of life, including education, politics, economy, and so on. The author overviews policies and recommendations with regard to GIA in the European context and demonstrates the tool developed within the RESET project – a checklist consisting of a set of questions about an intersectional gender perspective in three areas: excellence, implementation, and impact. The final chapter in this section by Tuija Lämsä offers a comprehensive overview of the types, diversified roles, and possible impact of CoPs (Communities of Practice) in organisations, based on the example of HEI. CoPs can be defined as a constellation of communities, consisting of various stakeholders, practitioners, experts, actors, and supporters who share common interests (for example, in gender studies or, more broadly, gender equality and DEI issues), knowledge, expertise, and know-how, and who want to exchange ideas (often across scientific disciplines), learn new practices and explore them in their daily work (as researchers, teachers, managers, project leaders, etc.). CoPs, whose members can be called gender change agents in HEIs, are analysed here through the example of communities created within, and hopefully beyond, the RESET project.

# 9

# FROM TRAINING TO ORGANISATIONAL CHANGE: THE ROLE OF STAKEHOLDER ENGAGEMENT IN EDUCATION ON GENDER EQUALITY AND DIVERSITY IN HIGHER EDUCATION INSTITUTIONS

*Agata Rudnicka and Aleksandra Różalska*

## Introduction

Each university has its own ecosystem of various stakeholders, processes, and resources, which translates into different expectations and the need to reconcile multiple interests. Academia consists of heterogeneous groups in terms of gender identity, age, sexual orientation, scientific discipline, position within the organisation, etc. Diversity of opinions, cultural values, beliefs, and stereotypes often make it difficult to discuss gender equality or, more broadly, the equal treatment of disadvantaged groups. Among the factors that can help universities to create friendly working and learning conditions that respect the multiple identities of each individual are equality in education and free access to training.

Education for equality and diversity is an issue that is contemporarily addressed by global and European institutions. The European Union emphasises the value of education for equality as a catalyst for the development of a society in which everyone enjoys their full rights (Drlel et al., 2023; A Union of Equality, 2020). The OECD highlights the value of gender equality in an ageing global society (Joining forces for gender equality, 2023), while UNESCO has developed a strategy for equality through and in education (From access to empowerment, 2019). Education covers multidimensional approaches related to fundamental human rights as it supports the process of equalising opportunities and increasing the representation of people at risk of exclusion, takes into account different perspectives as well as helps to build consensus and understanding of equality and diversity issues. Education makes the academic community more aware of the challenges and more knowledgeable about how to approach problem-solving. It facilitates the implementation and enforcement of common standards of behaviour.

DOI: 10.4324/9781003564348-16

In order to be effective and efficient, the process of education requires the convergence of several perspectives and approaches. In this chapter, we offer stakeholder theory (Freeman, 2010) as, in our opinion, it helps to map those groups that are essential for the fulfilment of the adopted educational goals. Methodologically, we refer to Bloom's Taxonomy (Shibaura, 2022), which facilitates the ordering of goals to be achieved in the area of inclusion and diversity, and to Kolb's Cycle (Kolb, 2014), which provides an opportunity for not only gaining theoretical knowledge but also experience, which stimulates sensitivity and provides a deeper understanding of equality and diversity.

## Stakeholders in the training system

### *Stakeholder theory*

The definition of a stakeholder, according to R. Edward Freeman, is "any group or individual who can affect or is affected by the achievement of the firm's objectives" and mission (Freeman, 1984, p. 25). Freeman, who is the advocate of stakeholder theory, insists that there is "the need to analyse stakeholder issues on three levels: rational, process and transactional" which translates into "mapping stakeholders, understanding organizational processes, and analysing interactions with stakeholders" (Freeman, 1984, p. 2). The first assumes identifying stakeholders in the organisation and their dominant perspectives thereon, the second demands investigating the organisation's possible relationships with these stakeholders and the ways they can be managed, and the third involves understanding "the set of transactions and bargains among the organisation and its stakeholders" (Freeman, 1984, p. 53) and how they can contribute to institutional processes.

Freeman and other scholars using stakeholder theory (Parmar et al., 2010; Simon, 2016) elaborate on the following concepts that are of interest to this chapter: heterogeneity of stakeholder groups, responsibilities of managers in balancing the needs and claims of various stakeholders, and the social responsibility of the organisation as its ethical obligation. In this context, Simon contends that

> a healthy company never loses sight of everyone involved in its success. Stakeholder theory says that if it treats its employees badly, a company will eventually fail. . . . A company can't ignore any of its stakeholders and truly succeed. There might be short-term profits, but as stakeholders become dissatisfied, and feel let down, the company cannot survive.
>
> *(Simon, 2016, n.p.)*

This demand of acknowledging the needs of multiple groups in the organisation puts a great deal of responsibilities on managers, who should be able to identify the needs of stakeholders, understand their concerns, and maintain an effective relationship with them, as well as have the skills and knowledge to respond to their expectations (Freeman, 1984, p. 26). These processes are seemingly uncomplicated, however, they require complex analyses, as well as the ability to deal with resistances and with the often conflicting demands of various internal stakeholders (Freeman, 1984, p. 246).

One needs to have these challenges and complex tasks in mind when identifying stakeholders in HEIs and recognising their needs and concerns.

### *Possible applications of stakeholder theory in HEIs in the context of training*

There is a growing interest in the application of the stakeholder theory in HEIs (Syed et al., 2023; Rumambi et al., 2019). As de Freitas Langrafe et al. (2020, p. 298) contend in their study, the following principles of stakeholder theory enrich HEIs and create a more inclusive and effective organisation:

> the active involvement of stakeholder groups in the decision-making process, the exchange of information about stakeholders' demands and preferences, the development of a relationship of mutual trust between stakeholders and the organization, and the inclusion of stakeholders in the strategic planning process.

Their study does not specifically mention training on diversity; however, in this chapter, we argue that this is a crucial aspect of building fruitful relationships with stakeholders. Such a perspective allows for building an "inclusive, intergenerational, anti-oppressive, and emancipatory" (Woodley & Rice, 2022) education on all levels of the training-learning process – from the design and review of the content of existing developmental activities/workshops and recognising stakeholders' needs, through delivery of training in the classroom and evaluation of the wider organisational context in which training is situated (Kapilashrami, 2021).

In the training systems implemented in HEIs, regardless of whether the training is offered as part of formal, informal, or non-formal education, different stakeholders can be identified (with particular attention to such disadvantaged groups as LGBTQIA+ communities, people with disabilities, and racial, ethnic, national, and religious minorities). The training process itself is also multi-directional. It should include both general topics (such as, for instance, mechanisms to counteract discrimination, different forms of bias and exclusion, overlapping axes of discrimination and power relations in HEIs, gender mainstreaming, and tailored-made actions) and particular themes related to the experiences of concrete stakeholders. Thus, several types of stakeholders should be recognised within the academic community, and one has to remember that this group is not homogenous as far as training needs are concerned, so they require training that tackles a diversity of themes and problems. In consequence, they acquire specific skills and develop gender-sensitive attitudes that are later helpful in their everyday activities, leading to a more efficient pursuit of their individual professional goals. Importantly, once adequately trained, the representatives of this group become informal trainers themselves who – in their particular work environments and through specific research and educational practices – effectively pass the acquired knowledge on to other employees and students.

Importantly, the group of stakeholders in HEIs is not limited exclusively to academic environments; it also comprises some external actors: legislative, funding, and training institutions and bodies (e.g., Ministry of Higher Education), which may include, for instance, representatives of non-governmental organisations, business, and public administration.

For the purpose of analysing the training needs of stakeholders and their demands for gender equality and diversity-related educational activities, in RESET, we applied a top-down approach to recognise those groups that play a key role in rooting for GE topics in HEIs.

Generally speaking, there are international and national groups of institutional stakeholders, such as the EU or national governmental agencies with many responsibilities and competencies in the field of gender equality development, including funding and legislation. However, our focus in this chapter is on the institutional stakeholders that have been included in the training processes undertaken within the RESET project so far.

Stakeholders within HEIs are crucial for the success of the training activities, hence, their needs must be carefully investigated and addressed. Importantly, they cannot be understood as a homogeneous group, which is why different competencies and training needs can pertain to the different groups of employees. For each group of employees, different forms of activities and engaging different dimensions of the gender equality issue may play an important role in the process of awareness-raising and the development of gender sensitivity. While managers typically determine the general culture (of, for example, inclusivity) within the institutions, they also remain responsible for the implementation of specific legal regulations and are accountable for the provision of support to lower level management as well as for all research and didactic processes taking place within the institution. Also, for these reasons, the improvement of competencies and the expansion of gender-sensitive knowledge remain crucial in building academic cultures of equality and spreading good practices within the whole community.

Equally important, though in a different regard, is the group of researchers, teachers, early stage researchers, and doctoral students, as they may trigger the dissemination of a gender-sensitive culture of inclusivity and diversity among community members – the research teams, departments, or groups of students. In this sense, they have a pivotal role to play as both receivers of knowledge and the transmitters of the desired behaviours, norms, and attitudes to other members of the community, including students. This group is also responsible for spreading norms in interpersonal relations, contacts, and daily communications. The knowledge, skills, and attitudes university teachers acquire should be transferred to educational processes so that students can experience diversity in their daily activities (Degtjarjova et al., 2018; Lau, 2014). Thanks to this, graduates will be more conscious citizens and employees who treat gender equality and diversity as a standard not as an exception. Students play a dual role in this ecosystem: they are beneficiaries who participate in training and people who see the impact of training – changes in organisational culture into increased inclusivity with a decreasing risk of discrimination.

Hence, targeting gender mainstreaming and gender-awareness training at particular stakeholder groups is of crucial importance, and this can be successfully achieved only through a thorough investigation of their training needs, demands, and knowledge gaps.

## Investigating the needs of various stakeholders in different HEIs: the RESET approach

### *The Stakeholder Engagement Standard*

The purpose of the Stakeholder Engagement Standard – developed by Accountability, which is a global research and consulting organisation providing innovative solutions to timely and important challenges according to social responsibility and sustainable development values – is "a generally applicable, open-source framework for assessing, designing, implementing and communicating an integrated approach to stakeholder engagement" (Misser et al., 2015, p. xx). These goals were used in the RESET project to establish the training needs of various academic groups. In particular, this tool helps to answer some basic questions with regard to stakeholders' roles and responsibilities in the organisation, such as the aim, scope, commitment, and level of engagement of stakeholders; the ways of integrating their involvement in governance, decision-making, strategy planning, and designing of intersectional training; and means to ensure the quality of the relationship between an HEI and its stakeholders and inclusive participatory practices (Misser et al., 2015). While identifying stakeholders and their demands in an HEI, one needs to take into consideration the following notions: dependency, responsibility, tension, influence and the diverse perspectives of respective stakeholders while planning the organisation's activities, decision-making processes, and training sessions, as well as the strategies of involvement of various groups in the HEI's operation (Misser et al., 2015).

### *Defining needs and problems to build effective training scheme*

The basis of a well-planned training programme is the examination of needs and the identification of problems, always having in mind intersectional (Woods et al., 2022) and co-design approaches. The map of the current situation (in the form of, for example, a checklist or questionnaire) provides concrete data regarding where the gaps are, and which areas require improvement. At the next level, much effort is put into developing knowledge and skills through a variety of training forms and tools, manuals, and other materials that can support the community. At the following stage, monitoring and evaluation is of great importance. In this phase, there are indicators and measures developed earlier that need to be monitored. There is also space to check to what extent the acquired knowledge and skills are preserved and whether further training is required. This stage remains unending under the continuous improvement approach. For an organisation with an implemented GEP, monitoring and improvement activities are a constant element of building excellence. All stages are accompanied by work related to the rooting of gender equality issues in the culture of the organisation. In addition to training activities addressed to specific groups and with specific goals, another important element is the participation of all interested parties in designing and developing capacity, which happens simultaneously with other activities.

Any training programme should start with a careful investigation of training needs and a detailed description of the educational goals. Delineating objectives remains an integral part of every training process. At this stage, a comprehensive analysis should differentiate between knowledge-related and skills-related goals. While the first part aims at the facilitation of understanding and remembering the concepts, approaches, and definitions, the second allows this knowledge to be used in practice.

The RESET project investigated the needs and requirements of four HEIs (from Poland, Portugal, Greece, and France) by analysing the condition of training activities with regard to GE and intersectionality (as of 2021), by investigating data obtained from a large-scale study conducted in 2021, and by interviewing representatives of different stakeholders from the four HEIs.[1] These complex inquiries facilitated the assessment of the number of people who participated in the GE and diversity training and whether they find such educational tools relevant. Additionally, the survey included questions on the preferred topics, formats, and modalities of training, as well as which target groups should be addressed while designing the training scheme.

During interviews with representatives from the four HEIs, several issues were brought up with regard to the specific training needs of each university. There were three important elements to this study:

1 Training needs in the context of national and regional organisations providing GE training to academic communities (for each partner, a map of stakeholders was worked out together, indicating each stakeholder's responsibility for and role in satisfying the training needs).
2 Training needs with regard to particular areas and topics related to GE and intersectionality. Specifically, it was discussed which topics are already covered by the existing training initiatives and which aspects of GE/diversity tend to be neglected or absent from the academic training offered by each university

internally. The participants were asked to assess which topics related to GE are still needed to support structural change and GEP implementation.

3 Training design, that is, who should be the primary target or priority group of training initiatives.

Furthermore, the participants were asked how to convince priority groups to take part in various forms of training and how to design training sessions with regard to time, format, methods used, etc.

### *RESET training activities: identification of needs*

One of the questions asked during the needs identification process of four HEIs in the RESET project tackled missing topics in the current professional practice – stakeholders were asked to self-assess which themes related to gender equality and diversity should be addressed. All four universities wanted to include training on gender and diversity dimensions as well as intersectionality in research and teaching, as it is one of the key requirements for successful GEP implementation and monitoring, building research teams, and preparing research projects in gender-inclusive, intersectional way. Some universities already started discussing this problem, while in others this aspect was considered unimportant or even "invented" by the EU. Three out of four HEIs pointed to gender-neutral and gender-inclusive language and communication as one of the issues that should be urgently addressed, for example, through institutional Gender Inclusive Language (GIL) recommendations.

University of Lodz (UL) wanted to have more training on mobbing and sexual harassment, while University of Bordeaux (UBx) – where the topic of sexual harassment is already mainstreamed – wanted to include a more detailed information about sexual harassment policies. University of Porto (U.Porto) and Aristotle University in Thessaloniki (AUTh) underlined the need to introduce knowledge on horizontal and vertical discrimination at the university, its consequences for all groups of employees, and possible measures to mitigate different forms of bias. All four universities found challenges in work–life balance (WLB) to be important, especially its gender implications and impact on female scientists. Similarly, all HEIs considered data on gender (in)equalities, gender-based violence, and intersecting systems of exclusion important. It seems crucial for all universities to explain during training such notions as prejudice, stereotypes, intersectionality, and different forms of discrimination, as well as mutual interrelationships between those concepts and the consequences they generate.

UL and UBx pointed out that members of the academic community often lack the skills to properly react to acts of discrimination and improper behaviour. This occurs not only due to the lack of abilities and/or knowledge but also due to an absence of formal measures, such as positive discrimination tools, anti-discrimination procedures, and awareness-raising activities. There was a general agreement among all four HEIs that practical training on how to react to discrimination in an everyday professional setting should be directed especially at the top and middle management. In addition, they insisted on focusing on gender bias and stereotypes in the teaching process and expressed the need for training on gender-sensitive attitudes and behaviours in the classroom and during interactions with students. Finally, unconscious prejudice and microaggressions analysed through gendered and intersectional lenses should also be included.

Importantly, there were some institution-specific topics. UL pointed out that training should tackle the lack of gender equality in academia as a systemic problem and give examples of concrete formal solutions to strengthen GE and diversity (for example, separate workshops on GEP and its importance for organisational culture). U.Porto wanted to include the aspect of gender dimension in decision-making processes and the issues of sexual orientation/identity/expression and possible anti-discrimination measures in this context. UBx observed the need to include more training for PhD students, especially for young female researchers and PhD candidates.

There were some differences among the four HEIs as far as priority target groups of training are concerned. Teacher/researcher respondents from Poland, Portugal, and Greece indicated that their own group should be trained first. In UL, respondents considered top and middle management as those who should be primarily trained and sensitised to different forms of discrimination. In U.Porto and AUTh, administrative staff, as well as middle and top management, were considered the most desired trainees of gender-related issues. Interestingly, it was suggested that for some groups (for instance, staff), such training should be mandatory. In AUTh, teachers/researchers and administrative staff were considered important to train first.

Identifying the needs for gender- and diversity-related training described earlier was essential for the development of the training programme that would be comprehensive, complex, and intersectional as well as adjusted to local contexts and various stakeholders.

## Designing a comprehensive, tailor-made, *à la carte* training programme

Achieving the intended educational goals with regard to gender and diversity requires conscious planning of methods and tools adapted to stakeholders and forms that allow them to meet different needs of learners. The content of the training scheme should also be considered to offer various stakeholders an opportunity to understand the importance of equality in their everyday work. This applies to both formal and non-formal education, although the former may have a greater potential, especially in higher education. As was described earlier, gender-sensitive education (Krišová & Polánková, 2020; Ostrouch-Kamińska & Vieira, 2016), the incorporation of a diversity perspective in research, and ensuring equal opportunities in all organisational processes are primary tasks for HEIs. Universities also play a crucial role in promoting DEI in society as a whole (Clauson & McKnight, 2018; Smith & Schonfeld, 2000). They are not only a place where knowledge is generated but also where social norms and behaviours that are important for creating a stable society are diffused and reinforced. Thanks to unlimited and free access to higher education, they are also a source of new values, which are developed and disseminated throughout society; as such, they have an important role to play in developing attitudes towards equality and diversity.

The academic community is made up of adults, so educational activities need to pay attention to the way this group learns (Zepke, 2005). The literature on adult education covers a wide range of approaches and strategies, which can be combined to achieve optimal results (Karge et al., 2011; Gouthro & Holloway, 2019; Parul et al., 2021; Sutton, 2022).

In the case of students, the inclusion of equality and diversity content can occur during the teaching process and be a natural part of it (Eden et al., 2024). However, for this to happen, teachers must have the necessary knowledge of intersectionality and the skills to embed it in their curricula and research. Another important aspect is creating conditions conducive to equality and diversity, that is, stimulating reactions to unethical phenomena (for example, to overlapping processes of discrimination) and developing desired behaviours. Therefore, education for equality uses many perspectives, approaches, and tools to achieve its different objectives (Corsino & Fuller, 2021; Mujtaba & Mujtaba, 2011).

Aligning the stakeholders' needs and expectations with the RESET project's goals was possible thanks to the use of different educational schemes. At the level of goal formulation, Bloom's Taxonomy of Goals was used (Grebin et al., 2020; Momen et al., 2022) to ensure that participants have the opportunity to acquire comprehensive knowledge and skills. In addition, the specificity of some topics made it possible to adapt elements of Kolb's Cycle (Kolb, 2014).

Bloom's Taxonomy was used as a supporting approach to setting training objectives, which helped design a series of courses focused on understanding and practising inclusive and anti-discriminatory behaviours. This tool also facilitated the introduction of an innovative training process, setting goals to simultaneous improvement of participants' knowledge and skills, and the development of coherent methods for achieving the planned outcomes. Recognising the importance of gender equality and intersectionality, and different perceptions of it, the training objectives included providing basic knowledge by explaining key issues and developing skills through practical activities, such as discussion, self-reflection, and critical evaluation. Table 9.1 outlines the educational purposes for gender equality in terms of Bloom's Taxonomy.

The diversity of educational aims was also reflected in the final version of the modules, which included both introductory training (mainly focused on knowledge acquisition) and in-depth training that provided opportunities to apply knowledge

**TABLE 9.1** Goals of gender equality education covering Bloom's Taxonomy

| *Element of Bloom's Taxonomy* | *Goal of gender equality education* |
|---|---|
| Remember | • Recollection of specific facts about gender equality.<br>• The list of definitions and concepts related to gender equality and diversity. |
| Understand | • Awareness about the issue of gender equality and its role at HEIs.<br>• Recognition of problems related to inequalities and improper behaviours.<br>• Identification of challenges.<br>• Discussion about the consequences of improper behaviour for individuals and institutions. |
| Apply | • Interpretation of observed behaviours and situations.<br>• Implementation of possible solutions and ideas oriented on prevention and enhancement of inclusive behaviours.<br>• The use of the knowledge to improve daily activities.<br>• Finding and applying solutions supporting gender equality and diversity.<br>• Demonstration of desired behaviours. |
| Analyse | • Comparison of different approaches towards gender equality and diversity.<br>• Examination of possibilities to implement different strategies. |
| Evaluate | • Assessment of applied solutions.<br>• Support for people with feedback about their approaches and practices.<br>• Selection of optimal options. |
| Create | • Designing new rules for all stakeholders.<br>• Development of awareness and skills helpful for identification and reaction to improper behaviours.<br>• Formulation of formal principles oriented on gender equality and inclusivity.<br>• Investigation of new ways of acting. |

*Source: Own elaboration*

through practical tasks. The final design of the training programme was preceded by extensive analysis and consultation (as described in previous parts of this chapter). One such activity was the analysis of existing training materials from similar projects. The study revealed a gap in training activities related to the psychological dimension, which is outside the equality education framework but still very much related to it. As a result, the training toolbox included topics such as setting personal boundaries, dealing with conflict, and managing difficult situations. Another key element was the involvement of different stakeholder groups to identify their training needs. As elaborated earlier, stakeholders were asked about their previous training experiences, preferred methods, and topics they would like to develop during the training sessions. Training themes were also identified based on data provided by partners and information gathered during the research, which highlighted gaps and challenges faced by individual universities. As a result, eight training modules were developed[2]:

1 Psychosocial risks at work – aggression, mobbing, and sexual harassment.
2 Reacting to harassment and inappropriate behaviours.
3 Diversity and inclusivity.
4 Enhancing diversity and inclusivity culture; preventing discrimination and unconscious bias – orientation on knowledge and skills competencies improvement.
5 Building positive relationships and enhancing positive attitudes towards diversity at work.
6 Work–life balance – orientation on knowledge competencies improvement.
7 GEP implementation.
8 Gender and diversity dimensions in research and teaching (including intersectionality and GIA).

In order to maintain consistency between the modules, a material planning scheme was developed that took into account both the technical and substantive aspects of the training process.

Each module focused on a different aspect of diversity and equality; however, the common feature was the complex interrelationship between gender and intersectionality. Each topic was described in the curriculum to facilitate navigating the toolbox. In addition to the elements listed earlier, each material also included a set of key concepts, definitions, and supplementary literature.

Considering the diversity of stakeholders, the modules included topics related to management, culture, intersectional academia, and psychology. The intention was to design a comprehensive solution where educational experiences would facilitate the practical application of knowledge. Hence, the emphasis on Kolb's Cycle, which aims to develop knowledge and skills by experiencing and learning about challenging situations and acquiring the ability to respond to them.

The need to integrate different dimensions of diversity and inclusivity is evident in various organisational processes. Training appears essential for proper understanding, stimulating cognitive curiosity and revealing aspects that are often overlooked in daily work, such as unconscious bias, microaggressions, or the risk of exclusion of certain stakeholder groups. Recognising inappropriate behaviour and being prepared to address it within the community is crucial. This responsibility lies with those who manage institutional processes, ensure safe working conditions, and respect the dignity of each individual within academia.

The expected societal change in thinking about equality and diversity will gradually occur at different levels. Training aims to ensure that different stakeholders are provided with the necessary knowledge and guidance to develop skills for practical

**TABLE 9.2** Designing training curriculum and materials process

| *Guiding questions* | *Planned element* |
|---|---|
| What? | • Goals the training should achieve.<br>• Processes it should cover (research, teaching, and administrative). |
| Who?<br>What are the main groups of stakeholders? | • Top and middle management.<br>• Academic staff (teachers, researchers, technicians, etc.).<br>• Administrative staff.<br>• PhD students.<br>• Students. |
| How?<br>Which form of training is the most appropriate? | • Online and face-to-face training.<br>• Individual and group work.<br>• Synchronous and asynchronous. |
| How?<br>What kind of methods could be used? | • Lecture<br>• Discussion<br>• Bubble dialog<br>• Tool coaching<br>• Action learning<br>• Design thinking<br>• Practical tasks<br>• Videos<br>• Self-reflection and individual work |
| How?<br>What kind of material does it contain? | • Scenario<br>• Template<br>• Checklist<br>• Test<br>• Presentation |
| How?<br>What competencies does it cover? | • Understanding the issues of gender equality and diversity.<br>• Approaching gender discrimination through intersectionality.<br>• Critical thinking on HEI's processes from an intersectional perspective.<br>• Reflection on improper behaviour in academia.<br>• Strategic thinking about improving inclusive and gender-sensitive processes in academia.<br>• Improving work–life balance mechanisms at institutional and individual levels.<br>• Designing anti-discrimination strategies and mechanisms at individual level.<br>• Preventing biases and prejudice, etc. |

*Source: Own elaboration*

application, for instance, in interpersonal relationships, bias-free decision-making, or improving the working environment. Each stakeholder group may have different needs regarding the purpose and scope of knowledge required to understand their current behaviour, identify areas for improvement, and ability to change practices that are not aligned with expected performance. Equally important is the exchange of experiences between different groups and the confrontation of their positions and viewpoints.

Training itself is only the first stage of a wider process that should take place formally, through new rules, documents, and procedures, and informally, through the lack of tolerance for unacceptable situations, behaviours, and attitudes. Participation in training helps shed light on different areas and raise awareness of issues essential for creating inclusive working and learning conditions.

## Conclusions (recommendations and lessons learnt)

The university ecosystem is made up of many stakeholders with diverse training needs and expectations in relation to gender equality and diversity; they might also be difficult to reach and challenging to encourage to take part in GE-related educational activities. Even a group as seemingly similar as academic staff may require different methods and forms of teaching, as they represent multiple disciplines and have unique experiences and approaches to gender issues. In RESET, we observed some reluctance towards gender and intersectionality training, especially among researchers and teachers from STEM, medicine, or other experimental sciences. Many people believe that such training is unnecessary and invented for the purpose of EU requirements. Some scholars believe that including gender and intersectionality in research and teaching is not relevant to their scientific activities, courses, projects, etc. The situation is similar when it comes to students' sensitivity, awareness, and openness. Adapting methods and tools for gender equality and inclusion training in higher education can therefore be challenging. Importantly, as ex-ante evaluations conducted in RESET show, the level of satisfaction from GE and diversity training is high and the more people participate therein, the more potential it has to change mindset and contribute to a more inclusive institutional culture.[3]

The approach implemented in the RESET project used a comprehensive set of tools, combining methods derived from management studies, psychology, communication studies, and sociology, enabling both the expansion of knowledge and the improvement of competencies among different stakeholders. Thanks to such designed training (i.e., using a variety of forms, formats, approaches, theories, methodologies, time frames, etc.), the process of building inclusive academia began through recognising diversity within diversity, addressing systemic inequalities and advocating for creating inclusive curricula and study programmes (Bhosale, 2024). The topics of training sessions were adjusted to particular academic groups, taking into consideration their heterogeneity and diversified experiences of power relations, discrimination, and exclusion. Some activities were designed for the broader community, during which synergy and learning about different perspectives on diversity, unconscious bias, or work–life imbalance were essential. Others were organised for specific types of stakeholders, such as managers or students, to discuss issues that are timely and important for these respective groups. People who participate in a training system designed in this way can attend sessions that increase their knowledge about particular topics or support them in taking roles of agents of institutional change if they are already prepared for it. The politics of training should, nonetheless, always take into consideration possible resistance (at both institutional and individual levels) and remain vigilant to such phenomena in HEIs as tokenism and surface-level diversity and representation, as well as limited intersectional research (Bhosale, 2024).

## Notes

1 Eleven stakeholders from four HEIs, representing teachers, researchers, middle management, and administrative staff, were interviewed in 2021. Then, in 2022, the topics of training were selected after each RESET partner consulted them with their institutional stakeholders and CoPs (e.g., Gender Equality Boards).
2 The training toolbox developed within RESET is available at https://toolkit.wereset.eu/#/training-toolbox/overview.
3 In the RESET project, we trained approx. 12,000 people representing the following academic stakeholders: students, PhD candidates, teachers, researchers, middle and top management, and administrative staff, including HR departments (Różalska et al., 2024).

## References


Bhosale, Uttkarsha. (2024). *Intersectionality in academia: Dealing with diverse perspectives*. https://www.enago.com/academy/intersectionality-in-academia/.

Clauson, Caitlyn, & McKnight, John. (2018). Welcome to campus: Planning for diversity, inclusion, and equity. *Planning for Higher Education*, *47*(1), 39–48. http://www1.scup.org/PHE/FMPro?-db=phe.fp5&-Lay=Home&-Format=Home.htm&-FindAny.

Corsino, Leonor, & Fuller, Anthony T. (2021). Educating for diversity, equity, and inclusion: A review of commonly used educational approaches. *Journal of Clinical and Translational Science*, *5*(1), e169. https://doi.org/10.1017/cts.2021.834.

de Freitas Langrafe, Taiguara, et al. (2020). A stakeholder theory approach to creating value in higher education institutions. *The Bottom Line*, *33*(4), 297–313. https://doi.org/10.1108/BL-03-2020-0021.

Degtjarjova, Irina, Lapina, Inga, & Freidenfelds, Davis. (2018). Student as stakeholder: 'Voice of customer' in higher education quality development. *Marketing and Management of Innovations*, *2*, 388–398. https://doi.org/10.21272/mmi.2018.2-30.

Driel, Barry, Donlevy, Vicki, & Melstveit Roseme, Maria. (Eds.). (2023). *Issue paper on gender equality in and through education – working group on equality and values in education and training. European Commission: Directorate-general for education, youth, sport and culture*. Publications Office of the European Union. https://data.europa.eu/doi/10.2766/915001.

Eden, Chima Abimbola, Chisom, Onyebuchi Nneamaka, & Adeniyi, Idowu Sulaimon. (2024). Cultural competence in education: Strategies for fostering inclusivity and diversity awareness. *International Journal of Applied Research in Social Sciences*, *6*(3), 383–392. https://doi.org/10.51594/ijarss.v6i3.895.

Freeman, R. Edward. (1984). *Strategic management. A stakeholder approach*. Pitman.

Freeman, R. Edward. (2010). *Stakeholder theory: The state of the art*. Cambridge University Press.

From access to empowerment: UNESCO strategy for gender equality in and through education 2019–2025. (2019). *UNESCO*. https://doi.org/10.54675/IJGQ3826.

Gouthro, Patricia A., & Holloway, Susan M. (2019). Using a multiliteracies approach to foster critical and creative pedagogies for adult learners. *Journal of Adult and Continuing Education*, *26*(2), 203–220. https://doi.org/10.1177/1477971420913912.

Grebin, Nataliya, Grabovska, Sophia, Karkovska, Ruslana, & Vovk, Anna. (2020). Applying Benjamin Bloom's taxonomy ideas in adult learning. *The Journal of Education, Culture, and Society*, *1*, 61–72. https://doi.org/10.15503/jecs2020.1.61.72.

*Joining forces for gender equality: What is holding us back?* (2023). OECD Publishing, https://doi.org/10.1787/67d48024-en.

Kapilashrami, Anuj. (2021). *Intersectionalityinformed framework for tackling racism and embedding inclusion and diversity in teaching & learning. A conceptual framework.* School of Heath and Social Care, University of Essex. https://s3.eu-west-2.amazonaws.com/assets.creode.advancehe-document-manager/documents/advance-he/2021%2004%2012%20Intersectional%20EDI%20framework%20for%20HE_1618994584.pdf.

Karge, Belinda Dunnick, Phillips, Kathleen M., Jesse, Tammy, & McCabe, Marjorie. (2011). Effective strategies for engaging adult learners. *Journal of College Teaching & Learning*, *26*. https://doi.org/10.19030/TLC.V8I12.6621.

Kolb, David A. (2014). *Experiential learning: Experience as the source of learning and development*. FT Press.

Krišová, Dagmar, & Polánková, Lenka. (Eds.). (2020). *Towards gender-sensitive education. Handbook for teacher trainers*. https://gendersensed.eu/wp-content/uploads/2020/08/A-Handbook-for-Teacher-Trainers.pdf.

Lau, Joseph W. C. (2014). Enriching stakeholder theory: Student identity of higher education. *American Journal of Industrial and Business Management*, *4*(12), 762–766. https://doi.org/10.4236/ajibm.2014.412082.

Misser, Sunil A., et al. (2015). AA1000 stakeholder engagement standard (SES). *Account Ability*. https://www.accountability.org/static/940dc017198458fed647f73ad5d47a95/aa1000ses_2015.pdf.

Momen, Abdu, Ebrahimi, Mansoureh, & Hassan, Ahmand M. (2023). Importance and implications of theory of Bloom's taxonomy in different fields of education. In Mohammed A. Al-Sharafi, Mostafa Al-Emran, Mohammed N. Al-Kabi, & Khaled Shaalan (Eds.), *Proceedings of the 2nd international conference on emerging technologies and intelligent systems, ICETIS 2022. Lecture notes in networks and systems* (Vol. 573). Cham: Springer. https://doi.org/10.1007/978-3-031-20429-6_47.

Mujtaba, Bahaudin G., & Lisa, Mujtaba Lisa. (2011). Diversity awareness and management in adult education. *Journal of College Teaching & Learning*, *24*. https://doi.org/10.19030/TLC.V1I3.1927.

Ostrouch-Kamińska, Joanna, & Vieira, Cristina C. (2016). Gender sensitive adult education: Critical perspective. *Revista Portuguesa De Pedagogia*, *50*(1), 37–55. https://doi.org/10.14195/1647–8614_50–1_2.

Parmar, Bidhan L., et al. (2010). *Stakeholder theory: The state of the art, publications management*. University of Richmond Scholarship Repository.

Parul, Sarin, J., Sheoran, P., & Phanden, R. K. (2021). Alliance in teaching-learning strategies and learning styles. In N. Kumar, S. Tibor, R. Sindhwani, J. Lee, & P. Srivastava (Eds.), *Advances in interdisciplinary engineering. Lecture notes in mechanical engineering*. Singapore: Springer. https://doi.org/10.1007/978-981-15-9956-9_7.

Różalska, Aleksandra, et al. (2024). *Demonstrations of the training programme in different partner universities*. RESET D4.4 Deliverable.

Rumambi, Hedy D., et al. (2019). Social responsibility of higher education institutions: A new insight into stakeholders. *GATR: Global Journal of Business and Social Science Review*, *7*(3), 185–193. https://doi.org/10.35609/gjbssr.2019.7.3(3).

Shibaura, Jessica. (2022). *Using Bloom's taxonomy to write effective learning outcomes*. https://tips.uark.edu/using-blooms-taxonomy/#gsc.tab=0.

Simon, Becky. (2016). *What is stakeholder theory and how does it impact an organization?* https://www.smartsheet.com/what-stakeholder-theory-and-how-does-it-impact-organization.

Smith, Daryl G., & Schonfeld, Natalie B. (2000). The benefits of diversity what the research tells us. *About Campus*, *5*(5), 16–23. https://doi.org/10.1177/108648220000500505.

Sutton, Halley. (2022). Adopt the best practices to meet unique needs of adult learners. *The Sucessful Registar*, *22*(8), 12. https://doi.org/10.1002/tsr.31033.

Syed, Raihan Taqui, et al. (2023). Higher education institutions and stakeholder analysis: Theoretical roots, development of themes and future research directions. *Industry and Higher Education*, *0*, 1–16. https://doi.org/10.1177/09504222231191730.

A Union of Equality: Gender Equality Strategy 2020–2025. (2020). *Communication from the Commission to the European Parliament, the Council, the European Economic and Social Committee and the Committee of the regions*. https://eur-lex.europa.eu/legal-content/EN/TXT/?uri=CELEX%3A52020DC0152.

Woodley, Xeturah, & Rice, Mary. (2022). *Designing intersectional online education critical teaching and learning practices*. Routlege.

Woods, Dorian R., Benschop, Yvonne, & van der Brink, Marieke. (2021). What is intersectional equality? A definition and goal of equality for organizations. *Gender, Work & Organization*, *29*, 92–109. https://doi.org/10.1111/gwao.12760.

Zepke, Nick. (2005). Diversity, adult education and the future: A tentative exploration. *International Journal of Lifelong Education*, *24*(2), 165–178. https://doi.org/10.1080/02601370500124181.

# 10

# THE ROLE AND ACTIONS OF A TRAINER IN CONDUCTING TRAINING AND WORKSHOPS ON DIVERSITY AND GENDER EQUALITY AT UNIVERSITIES

*Mateusz Hauk and Paulina Szymańska*

## Introduction

Diversity and equality training have gained significant popularity among many organisations, including the higher education sector (Holladay & Quiñones, 2008; Liberman et al., 2011). Research has demonstrated that companies with greater ethnic and gender diversity are more likely to outperform the national industry median in financial performance (Hunt et al., 2015) and report 19% higher innovation revenues (Lorenzo et al., 2018). Consequently, there is a growing interest in implementing diversity, equity, and inclusion (DEI) initiatives, including webinars, workshops, and training. Significant financial resources have been increasingly allocated to various DEI training programmes in recent years. By 2026, global spending on DEI-related initiatives is projected to exceed $15 billion (Global Industry Analysts Inc., 2021).

Companies recognise numerous benefits of these initiatives, such as fostering a positive organisational culture and common identity, reducing conflicts, minimising turnover, and enhancing employee efficiency and commitment, all of which contribute to various forms of profit for the company (Al-Badareen & Al-Ghazzawi, 2018; Devine & Ash, 2022; Zhang & McGuire, 2021). However, further research is needed, particularly concerning the long-term effects of DEI policies (Devine & Ash, 2022).

At the same time, research highlights the employees' perspective. For instance, 56% of personnel perceive diversity issues as generally positive, with women holding more favourable attitudes than men (Pew Research Center, 2023). Additionally, over three-quarters of job seekers and employees across various institutions indicate that diversity in the workplace plays a crucial role in recruitment, particularly for minority groups (Glassdoor, 2021). In practice, a company's approach to

DOI: 10.4324/9781003564348-17

cultivating and maintaining a DEI-friendly company culture influences how potential employees perceive the company, how it is evaluated, and whether candidates choose to apply for a job in this organisation.

Although DEI training has gained significant popularity, it is distinct from other developmental activities due to their focus on challenging social issues. They require participants to discuss difficult topics and aim to foster cultural change (Lindsay, 1994). Unlike other common training sessions, this training explicitly addresses attitudes, beliefs, and behaviours that can be deeply ingrained in socio-cultural context, very private and resistant to change (Holladay & Quiñones, 2008; Jeffries, 2023). Therefore, with regard to the effectiveness of DEI programmes, such factors as specific work environment, theoretical approaches to training (Bezrukova et al., 2012), as well as the support of management and sufficient financial resources to implement the programme may be crucial.

In addition, in public and non-profit sectors (including a large number of HEIs), a greater emphasis on social justice or learning opportunities has become increasingly important, rather than the potential material gain for the organisation (Bond & Haynes, 2014). While changing attitudes to diversity in academia, we should consider its hierarchical structure, attachment to tradition and high diversity of needs. Resistance to change may stem from long-standing cultural norms that sometimes marginalise or exclude some groups (particularly in certain disciplines, such as STEM). In addition, university culture emphasises autonomy, whereas some studies suggest (Dobbin & Kalev, 2018) that DEI training is often perceived as aiming to control employees' behaviour, which can affect its reception and impact.

## Various approaches to DEI training

According to a review by Wang et al. (2023), various theoretical frameworks have been applied so far in DEI training, such as Aversive Racism Theory (Gaertner & Dovidio, 1986), The Hopkin's Process Model of Ethics-Diversity Relationships (Hopkins, 1997), Bennett's Model of Developing Intercultural Sensitivity (Bennett, 1986), Self-categorization Theory (Turner et al., 1987), Social Learning Theory (Rotter, 1954), Contact Hypothesis (Allport, 1954), or Dissonance Theory (Festinger, 1957). The primary emphasis was placed on topics related to stereotyping, social categorisation, implicit bias, micro- and macro-aggressions, power, and health disparities. Some of the analysed DEI programmes also included building self-awareness (O'Connor et al., 2019), perspective-taking (Lindsey et al., 2015), empowerment, personality types (Hostager & De Meuse, 2008), emotional intelligence, and growth as well as fixed mindsets (Powell et al., 2021). Based on the information mentioned earlier, it is evident that the structure of DEI training first and foremost focuses on the dissemination of knowledge related to diversity, equity, and inclusion. Secondary emphasis is placed (though in our opinion still in an insufficient way) on psychological aspects of human functioning that might help to promote change.

Awareness and understanding of equality-related concepts are essential; however, if discussed in isolation, they appear insufficient. It is necessary to develop specific competencies that facilitate the practical application of this knowledge. Not only should training take into account issues related to demographic or cultural aspects but also the individual emotional state of employees and their ability to regulate in this area (Al-Badareen & Al-Ghazzawi, 2018). It should also take into consideration other circumstances, for instance, the role of communication (TAAZ Inc., n.d.).

## Training structure

As research shows (Dobbin & Kalev, 2016; The Arbinger Institute, 2023), the current process of diversity training implementation is not entirely conducive to effectiveness. Therefore, it is crucial to implement and improve the existing practices focused mostly on knowledge transfer and thus not only bring short-term measurable effects but also strengthen building diversity in this organisational culture, taking into account mutual trust, respect, and openness as well as facilitating effective communication.

DEI programmes should incorporate two essential aspects. First, the training structure should be grounded in a knowledge-based approach that enhances participants' awareness of the relevant issues; second, it should focus on developing particular competencies that enable a practical application of this knowledge. Next is an outline of the sample topics, divided into two sections: theoretical and psychological, based on experiences deriving from the RESET project (Różalska et al., 2022).

### *The theoretical DEI training section*

Based on the knowledge-attitude-behaviour model (Bettinghaus, 1986), there is a relationship between information acquisition and changes in viewpoint and a person's actions. As such, the trainer's primary objective in the first section is to convey comprehensive knowledge on diversity and inclusion within the organisation. This also entails familiarising participants with ethical principles and legal frameworks applicable at institutional, national, and global levels. Outlining the possibilities for incorporating DEI policies into academic and organisational practices (including the recruitment process, ongoing employment, as well as educational and research contexts) appears essential. Nair and Vohra (2015) have pointed out that introducing DEI principles may increase employees' satisfaction and engagement. Therefore, gaining knowledge of the company's approach to diversity and inclusivity may indirectly support their professional effectiveness. Another, but somehow complementary, aspect is the discussion on Gender Equality Plans and strategies for their effective implementation. Lastly, this section explores particular tools and methods that promote DEI in the workplace, emphasising the need to deepen understanding of causes and types of discrimination and the ability to recognise, identify, and apply potential solutions. This approach naturally aligns with the concept of an inclusive culture.

Following are some examples of topics and learning areas that could be covered in the theoretical section (based on the experiences of the RESET project [Różalska et al., 2022]):

- **Psychosocial risks at work – aggression, mobbing, and sexual harassment:** the analysis of different psychosocial risks at work, with a particular focus on aggression and violence (including mobbing and sexual harassment), definitions, and applicable typologies; an investigation of possible causes of aggressive behaviour within the work environment and its repercussions for individuals (including victims, perpetrators, and witnesses) as well as for the organisation as a whole; relevant legal regulations designed to protect victims of violence in the workplace.
- **Diversity and inclusivity:** ways of understanding diversity and inclusiveness, an intersectional approach as applied in research, teaching, and administration; the analysis of the current state of DEI in each organisation (e.g., the maturity model); ways to build a successful DEI strategy and to monitor its implementation; exemplary tools and creative ideas for building individual and group DEI awareness.
- **Preventing discrimination and unconscious bias:** different types of discrimination, stereotyping, and prejudice; their causes and consequences; gender-based discrimination and its types; the concept of unconscious biases; types of bias (e.g., perception bias, halo effect, and attribution bias) and their consequences.
- **Gender and diversity dimensions in research and teaching:** gender mainstreaming in the university context (research and teaching areas); diversity mainstreaming through an intersectional approach; guidelines for including gender and diversity dimensions in research and teaching practices; selected tools, such as Gender Impact Assessment (GIA).
- **Gender Equality Plan (GEP) implementation:** GEP's goals, design, and composition; stages of implementing GEP at universities and possible obstacles or resistances; the analysis of various stakeholders' primary responsibilities; communication and dissemination strategy to sensitise the community on critical areas of GEP; GEP evaluation and monitoring of its quality and impact.

### *The psychological DEI training section*

In the second part, which addresses psychological aspects, the trainer should emphasise the practical application of the tools and skills acquired during theoretical education. This approach helps to lay the groundwork for changing attitudes regarding diversity, equity, and inclusion and fostering an inclusive organisational culture in the long term. It is advisable to include aspects related to setting boundaries, assertiveness, and methods for taking care of one's psychological well-being. A glance at the literature indicates that skills and competencies, such as empathy, active listening, self-awareness, and the ability to relate to others, are key facilitators in shaping a culture of diversity and inclusivity within a company (Hays-Thomas et al., 2012). As evidence shows, effective, open, and respectful communication prevents misunderstandings and makes misleading assumptions as well as lowers the occurrence of conflicts (Bucăţa & Rizescu, 2017; Daya, 2014). Therefore, this part should focus on developing participants' interpersonal skills through honest, free-flowing communication and identifying the needs of individuals and the interests and needs of others working within the organisation. Accurate reading of incoming information and clearly expressing one's expectations is crucial in a diverse and socially demanding environment. Emphasis should also be placed on conflict management skills, such as those outlined in the theoretical framework of transactional analysis. Finally, attention should be given to work–life balance (WLB) and integrating personal and professional spheres of life. In line with previous studies, Beauregard et al. (2020) underline the connection between WLB and culture or gender, linking it with individual and social norms and values (e.g., division of domestic labour or living in multigenerational houses). Moreover, building employees' ability to combine work with private life effectively reflects the organisation's commitment to supporting diversity (Deloitte, 2012; Nair & Vohra, 2015).

Exemplary training areas in the psychological section are as follows (based on the experiences of the RESET project) (Różalska et al., 2022):

- **Reacting to harassment and inappropriate behaviours:** the concept of Nonviolent Communication (NVC) as a way of responding to inappropriate behaviour in the work environment; assertive setting of boundaries; building skills of giving feedback, including using I-message; analysis of one's own emotional experiences and self-regulation strategies; sources of external support.
- **Enhancing diversity and inclusivity culture:** the role of leaders and their dilemmas in building DEI culture in the academic environment; psychological mechanisms of communication; ways of describing expectations and concluding contracts with employees; analysis of recognition signs as necessary in building a favourable relationship with the superior and co-worker; feedback as a method of informing employees about expectations with regard to DEI; skills enforcement procedure; the leader's ability to draw formal and informal sanctions for non-compliance with the principles of DEI-related issues.
- **Building positive relationships and enhancing positive attitudes towards diversity at work:** various approaches to conflicts and triggers that increase tension in relationships; analysis of ways of resolving conflicts; main obstacles in communication and ways of reacting in difficult situations; explanation of the concept of ego states and transactions; analysis of the types of recognition signs and their impact on human functioning; stroke economy; interpersonal

games based on Transactional Analysis, including modes of time structuring, specificity of games, and the game sequence, most commonly played games, and potential benefits of engaging in games; drama triangle analysis.
- **Work–life balance:** different approaches to work–home relationship; JD-R model; individual and organisational consequences of work–life imbalance; application of diversity context to WLB in academia; tools for maintaining a healthy and satisfactory relationship between work and non-work areas, that is, contracting, assertive refusal and asking for help.

The training structure can be tailored and customised to fit the specific needs of a particular group or organisation. Simultaneously, it is crucial to incorporate both theoretical aspects of DEI and the development of psychosocial competencies that facilitate the practical application of the acquired knowledge.

### The role of a trainer

Based on the guiding principles by Pendry et al. (2007), Bezrukova et al. defined diversity training as "a distinct set of programs aimed at facilitating positive intergroup interactions, reducing prejudice and discrimination, and enhancing the skills, knowledge, and motivation of people to interact with diverse others" (2012, p. 208). Considering the specific nature of such training sessions, which differ from other types of workshops, it seems there are many challenges to address when conducting them. Among these, it is vital to highlight the need to overcome resistance, address sensitive and sometimes controversial topics, and create an atmosphere where participants feel safe to engage openly. This might make this type of training very intense, emotional, and uncomfortable for both participants and trainers, underscoring the pivotal and multidimensional role of the trainer, including such activities as dealing with resistance, facilitating, mediating, educating, and instructing (Morukian, 2022).

### Handling resistance

Effective DEI trainers should possess extensive knowledge of how to handle resistance constructively. Defiance can manifest in various forms, but in higher education, common concerns include a feeling of being overburdened, and contestation of the relevance of these topics to academics, who often perceive themselves as highly aware and educated. Hence, it is recommended that trainers acknowledge resistance and recognise and validate participants' concerns without being judgemental. This can be achieved by facilitating open discussions during which trainees can voice their apprehensions (Morukian, 2022).

While conducting DEI training, it is useful to focus on honing competencies and leading to a real change (in terms of attitudes and behaviours of participants as well as the whole organisation), which requires substantial effort from participants and is usually time-consuming. Adopting this approach and encouraging people to participate in extended development activities might lead to reluctance and resistance. It is important to keep in mind that this process is likely to trigger typical psychological reactions of defiance. Every change, whether welcome or not, involves certain stages that must be navigated. According to Kübler-Ross' Change Curve Model (Kübler-Ross, 2014; Savolainen, 2016), which has demonstrated its effectiveness in corporate and institutional settings, employees may experience five stages of personal transition: denial, anger, bargaining, depression, and acceptance. When encouraging individuals to adopt a more diversity- and equality-focused mindset, these phases of adapting to change are likely to emerge. Additionally, when implementing development initiatives and inviting participation, one might face denial, reluctance, and refusal – especially when the activities require time and effort. Following are some practical strategies for addressing resistance at the initial stage of the process of change adaptation:

- Highlight potential benefits coming from engagement in development activities.
- Illustrate potential risks of not participating in development activities and resisting ongoing changes.
- Emphasise the inevitability of change.
- Demonstrate the crucial role of participants in the development process and their ability to influence change.
- Ask about any challenges or obstacles that might hinder their full engagement in the development process and encourage them to come up with their own optimal solutions.

Naturally, for the interventions mentioned earlier, the trainer in higher education needs support from others who are involved in introducing changes to the work environment. These stakeholders include authorities who can lead by example and emphasise the significance of the programme, as well as communication departments, top management, and executives.

## Building a climate of psychological safety, facilitation, and mediation

One of the most important aims of a trainer is to encourage participants to speak openly and share their perspectives, which can enhance the perception and awareness of others. To achieve this, it is crucial to set the stage for deep and potentially difficult work, conversation, and exchange. Thus, the primary objective for trainers is to establish a foundation of trust and openness from the very beginning by ensuring psychological safety (Edmondson, 2019; Morukian, 2022). Therefore, the introductory phase of the training should focus on establishing certain foundational elements: setting clear goals that emphasise the importance of covering these topics and providing an overview of the training process along with the agenda.

Apart from that, it is extremely important to begin with an opening round to create a safe atmosphere and start building connections between participants. The next step is to co-design a mutual agreement or "contract", establishing a safe space where participants feel protected and valued (Jeffries, 2023). In our view, certain ground rules are essential for training sessions – using each other's names (to promote equality in relationships), maintaining confidentiality, staying "here and now", showing respect, refraining from judgement, and sharing responsibility for the outcomes of training.

Although some participants claim that such activities are time-consuming and unnecessary (as they are not directly related to the subject matter), we insist that setting the stage is very beneficial for reducing the possibility of potential conflicts. It becomes easier to discuss difficult matters and listen to each other attentively and with respect.

At the same time, facilitation is needed when addressing complex issues related to gender and diversity. Since every participant may bring something unique to group discussion and interaction – different perspectives, difficulties, misunderstandings, fears, and concerns – the trainer should moderate and mediate the process, encourage everyone to speak up, stimulate participants to share their thoughts, and simultaneously show empathy and understanding (Jeffries, 2023; Lindsay, 1994).

In doing so, it is extremely important to actively listen and encourage others to reveal their perspectives, which can be enriching for the whole group. Additionally, the trainer should model proper behaviour when discussing potentially difficult topics. It is recommended to keep the following points in mind:

- Give participants time and space to talk, and be willing to listen.
- Seek to understand and assist the other person, especially when they are formulating their thoughts.
- Focus your attention on the other person; avoid activities that put the spotlight on yourself or turn training into an entertaining event.
- Trust the other person to come up with solutions that work best for them, particularly when the task involves creating or figuring out new options.

In situations of tension between participants, a conflict resolution procedure is recommended (Gordon, 2011; see also, Thomas Gordon International: Diving into Conflict Resolution). The trainer's role is to moderate the conversation by adhering to the core principles of conflict resolution and assisting participants in listening to each other, especially when discussing interests and needs. This process can be enriching for others who take part in a training session and serve as an example of how to conduct difficult conversations constructively.

## Education, instruction, and learning process

As mentioned earlier, DEI trainers need to possess extensive knowledge of diversity and gender issues in order to be able to introduce these concepts in a way that is relatable to participants' experiences and a university setting (Lindsay, 1994). Importantly, trainers should adopt the role of educators rather than enforcers. As some comments or behaviours of the participants can be emotional triggers for the trainer and attendees, it is vital to balance empathy, compassion, and curiosity, avoiding persuasion and imposition (it is better to avoid the Rescuer or Persecutor role from the Drama Triangle) (Stewart & Joines, 2012). Encouraging critical thinking and self-reflection allows participants to draw their own conclusions about the importance of gender equality and diversity (Chrobot-Mason & Quiñones, 2002).

In order to enhance the odds of a successful training aimed at increasing awareness and knowledge among participants as well as creating the opportunity to work on real, difficult cases and hone their skills, it is essential to use Kolb's Experiential Learning Cycle (Kolb, 1984/2014; Morukian, 2022). This tool underpins many effective development and education strategies and is also commonly applied in business settings. Developed by Kolb, building on Lewin's T-groups observations (Highhouse, 2002), this model integrates two crucial learning processes: receiving information (through senses and emotions as well as symbol/language) and information processing (through introspection and reflection as well as action and activity). This framework of learning (Kolb, 1984/2014) assumes the following elements/stages of the educational process:

Trainers can facilitate effective learning through concrete experiences (incorporating activities and discussions that allow participants to share their own insights related to diversity and gender issues), reflective observation (facilitating sessions where participants reflect on what they learned), abstract conceptualisation (providing theoretical frameworks and models that help participants understand what has happened in the broader context of DEI), and active experimentation (encouraging

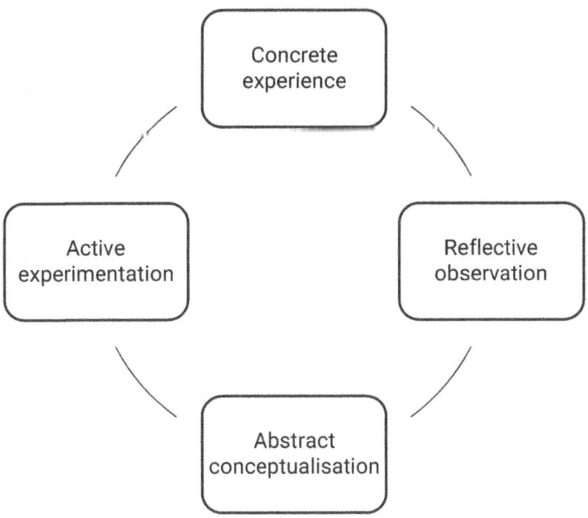


**FIGURE 10.1** Kolb's Cycle (Kolb, 1984/2014)

participants to apply what they have learned through role-playing, case studies and real-world applications).

A particularly useful method for preparing participants to handle difficult situations related to gender or diversity issues is tool coaching (Gut & Haman, 2014), which entails supporting trainees in dealing with real, difficult situations by providing detailed feedback from others and honing skills to cope with common challenges. The trainer facilitates the meeting by first gathering and selecting challenging real-life situations from the participants. Then the group examines them through role-playing, discussions, or interventions, collects feedback on effective and risky responses, and develops guidelines for handling similar situations. This includes advice on appropriate behaviour, the most desirable conversation structure, key points to be mindful of, and actions to avoid. Finally, participants practise a conversation algorithm, having been previously provided with theoretical background, in smaller groups.

In addition, useful methods for working on real situations can be derived from the action learning approach (Cho & Egan, 2009) or Balint groups (https://balint-society.org.uk/balint-groups-and-balint-method). To encourage people to work on their own difficult situations and provide them with the support from other trainees, fostering an atmosphere of mutual help and development, the trainer may organise part of the workshop in the following way:

- The participant presents the situation and formulates a question or dilemma to be handled.
- The group asks questions to clarify their understanding of the situation, with the participant providing answers.
- The group offers hints and suggestions, sharing their own perspectives and challenges, and generating a range of possible solutions, advice, and recommendations.

The participant reviews the received feedback and decides how and to what extent they wish to implement it. Another beneficial tool to consider to stimulate reflection and encourage work on difficult real-life situations is the Bubble Dialogue Method (Chomczyński, 2008). The main objective of this approach is to increase participants' knowledge on and awareness of the phenomena that are discussed, encourage them to exchange ideas, and find solutions to challenging situations. It consists of the following steps:

- The group work begins with showing a fragment of a video depicting particular scenes (e.g., aggression or psychological violence).
- The video serves as a point of departure for discussion among participants.
- It can also serve as a foundation for role-playing or in-depth discussions aiming to explore new behaviours or alternative responses to challenges. The following

questions can be asked at this stage: What should be done in such cases? How should one behave?

As elaborated earlier, there are many methods that enable trainers to work on real-life situations and provide participants with practical knowledge and skills useful when dealing with difficult scenarios. It is worth emphasising, however, that before using methods such as tool coaching, bubble dialogue, or others, building an atmosphere of openness is crucial.

## Conclusion

DEI training can be perceived as a specific form of development requiring a careful approach as well as diverse competencies and actions from trainers. This is especially true when training sessions are conducted at HEIs, with the aim of changing the organisational culture or enhancing a new, more inclusive mindset, which inevitably means longer developmental processes that demand substantial effort and engagement.

The role of a trainer in DEI training at universities is multifaceted, requiring a blend of knowledge, empathy, and mindful interventions. By creating a safe and open learning environment, thoughtfully addressing resistance and utilising Kolb's Cycle, trainers can foster a culture of inclusivity and respect. Therefore, it seems useful to employ engaging and "practical" forms and methods of work, such as tool coaching, among others.

A review of relevant literature, together with our own personal experiences related to implementing such projects, suggests that dissemination of knowledge related to diversity, equity, and inclusion is, of course, crucial in these developmental efforts, although insufficient on its own. It is crucial and highly recommended to focus on psychological aspects of human functioning, in order to equip participants with knowledge on and awareness of psychological mechanisms that influence people's behaviour. Moreover, it is equally important to provide knowledge on communication, relationship-building, conflict resolution dealing with challenging situations, and finally, to equip individuals with practical tools allowing them to build satisfying relationships while taking care of both themselves and others.

## References

Al-Badareen, Rokaya, & Al-Ghazzawi, Hadeel. (2018). Diversity training and its impact on the employees' organizational commitment: Job satisfaction as a moderating variable in the classic fashion apparel industry company. *International Journal of Business and Management, 13*(10), 116–128. https://doi.org/10.5539/ijbm.v13n10p116.

Allport, Gordon W. (1954). *The nature of prejudice*. Addison-Wesley.

The Arbinger Institute. (2023, August 8). *Why most DEI training programs don't work.* Arbinger Institute Blog. https://arbinger.com/blog/why-most-dei-training-programs-dont-work/.

Beauregard, T. Alexandra, Adamson, Maria, Kunter, Aylin, Miles, Lilian, & Roper, Ian. (2020). Diversity in the work-life interface: Introduction to the special issue. *Equality, Diversity and Inclusion: An International Journal, 39*(5), 465–478. https://doi.org/10.1108/EDI-04-2020-0097.

Bennett, Milton J. (1986). A developmental approach to training for intercultural sensitivity. *International Journal of Intercultural Relations, 10*(2), 179–196. https://doi.org/10.1016/0147–1767(86)90005–2.

Bettinghaus, Erwin P. (1986). Health promotion and the knowledge-attitude-behavior continuum. *Preventive Medicine, 15*(5), 475–491. https://doi.org/10.1016/0091–7435(86)90025–3.

Bezrukova, Katerina, Jehn, Karen A., & Spell, Chester S. (2012). Reviewing diversity training: Where we have been and where we should go. *Academy of Management Learning & Education, 11*(2), 207–227. https://doi.org/10.5465/amle.2008.0090.

Bond, Meg A., & Haynes, Michelle C. (2014). Workplace diversity: A social–ecological framework and policy implications. *Social Issues and Policy Review, 8*(1), 167–201. https://doi.org/10.1111/sipr.12005.

Bucăţa, George, & Rizescu, Alexandru Marius. (2017). The role of communication in enhancing work effectiveness of an organization. *Land Forces Academy Review, 22*(1). https://doi.org/10.1515/raft-2017–0008.

Cho, Yonjoo, & Egan, Toby Marshall. (2009). Action learning research: A systematic review and conceptual framework. *Human Resource Development Review, 8*(4), 431–462. https://doi.org/10.1177/1534484309345656.

Chomczyński, Piotr. (2008). *Mobbing w pracy z perspektywy interakcyjnej. Proces stawania się ofiarą [Mobbing at work from an interactive perspective. The process of becoming a victim].* Publishing house of the University of Lodz.

Chrobot-Mason, Donna, & Quiñones, Miguel A. (2002). Training for a diverse workplace. In K. Kraiger (Ed.), *Creating, implementing, and managing effective training and development*. San Francisco: Jossey-Bass.

Daya, Preeya. (2014). Diversity and inclusion in an emerging market context. *Equality, Diversity, and Inclusion: An International Journal, 33*(3), 293–308. https://doi.org/10.1108/EDI-10-2012-0087.

Deloitte. (2012). *Waiter, is that inclusion in my soup? A new recipe to improve business performance*. Deloitte Research Report. https://www2.deloitte.com/content/dam/Deloitte/au/Documents/human-capital/deloitte-au-hc-waiter-inclusion-soup–0412.pdf.

Devine, Patricia G., & Ash, Tory L. (2022). Diversity training goals, limitations, and promise: A review of the multidisciplinary literature. *Annual Review of Psychology, 73*, 403–429. https://doi.org/10.1146/annurev-psych-060221–122215.

Dobbin, Frank, & Kalev, Alexandra. (2016). Why diversity programs fail. *Harvard Business Review*. https://hbr.org/2016/07/why-diversity-programs-fail

Dobbin, Frank, & Kalev, Alexandra. (2018). *Why doesn't diversity training work? The challenge for industry and academia*. Harvard University. https://scholar.harvard.edu/files/dobbin/files/an2018.pdf.

Edmondson, Amy C. (1999). Psychological safety and learning behavior in work teams. *Administrative Science Quarterly, 44*(2), 350–383. https://doi.org/10.2307/2666999.

Festinger, Leon. (1957). *A theory of cognitive dissonance*. Stanford University Press.

Gaertner, Samuel L., & Dovidio, John F. (1986). The aversive form of racism. In J. F. Dovidio & S. L. Gaertner (Eds.), *Prejudice, discrimination, and racism*, 61–89. Academic Press.

Glassdoor. (2021). *What job seekers really think about your diversity and inclusion stats*. https://www.glassdoor.com/blog/diversity/.

Global Industry Analysts Inc. (2021). *With global spending projected to reach $15.4 billion by 2026, diversity, equity & inclusion takes the lead role in the creation of stronger businesses*. https://www.prnewswire.com/news-releases/with-global-spending-projected-to-reach-15–4-billion-by-2026-diversity-equity-inclusion-takes-the-lead-role-in-the-creation-of-stronger-businesses–301413808.html.

Gordon, Thomas. (2011). *Leader effectiveness training, L.E.T.* Tarcherperigree.
Gut, Jerzy, & Haman, Wojciech. (2014). *Psychologia szefa 2. Coaching narzędziowy [Boss psychology. Tool coaching]*. Helion Publishing House.
Hays-Thomas, Rosemary, Bowen, Alyinth, & Boudreaux, Megan. (2012). Skills for diversity and inclusion in organizations: A review and preliminary investigation. *The Psychologist-Manager Journal, 15*(2), 128–141. https://doi.org/10.1080/10887156.2012.676861.
Highhouse, Scott. (2002). A history of the T-group and its early applications in management development. *Group Dynamics: Theory, Research, and Practice*, *6*(4), 277–290. https://doi.org/10.1037/1089–2699.6.4.277.
Holladay, Courtney L., & Quiñones, Miguel A. (2008). The influence of training focus and trainer characteristics on diversity training effectiveness. *Academy of Management Learning & Education, 7*(3), 343–354.
Hopkins, Willie E. (1997). *Ethical dimensions of diversity*. Sage Publications.
Hostager, Todd J., & De Meuse, Kenneth P. (2008). The effects of a diversity learning experience on positive and negative diversity perceptions. *Journal of Business and Psychology, 23*(1), 127–139. https://psycnet.apa.org/doi/10.1007/s10869-008-9085-x.
Hunt, Dame Vivian, Layton, Dennis, & Prince, Sara. (2015). *Why diversity matters*. https://www.mckinsey.com/capabilities/people-and-organizational-performance/our-insights/why-diversity-matters.
Jeffries, D. J. (2023). Adult learning principles in corporate diversity, equity, and inclusion training. *Adult Learning, 34*(4), 256–259. https://doi.org/10.1177/10451595231184927.
Kolb, David A. (1984/2014). *Experiential learning: Experience as the source of learning and development*. FT Press.
Kübler-Ross, Elisabeth. (2014). *On death and dying: What the dying have to teach doctors, nurses, clergy and their own families*. Simon and Schuster.
Liberman, Benjamin E., Block, Caryn J., & Koch, Sandy M. (2011). Diversity trainer preconceptions: The effects of trainer race and gender on perceptions of diversity trainer effectiveness. *Basic and Applied Social Psychology, 33*(3), 279–293. https://doi.org/10.1080/01973533.2011.589327.
Lindsay, Cindy. (1994). Things that go wrong in diversity training: Conceptualization and change with ethnic identity models. *Journal of Organizational Change Management, 7*(6), 18–33. https://doi.org/10.1108/09534819410072683.
Lindsey, Alex P., King, Eden, Hebl, Michelle, & Levine, Noah. (2015). The impact of method, motivation, and empathy on diversity training effectiveness. *Journal of Business and Psychology, 30*(4), 605–619. https://psycnet.apa.org/doi/10.1007/s10869-014-9384-3.
Lorenzo, Rocio, Voigt, Nicole, Tsusaka, Miki, Krentz, Matt, & Abouzahr, Katie. (2018, January 23). *How diverse leadership teams boost innovation*. Boston Consulting Group. https://www.bcg.com/publications/2018/how-diverse-leadership-teams-boost-innovation
Morukian, Maria. (2022). *Diversity, equity, and inclusion for trainers: Fostering DEI in the workplace*. Association for Talent Development.
Nair, Nisha, & Vohra, Neharika. (2015). *Diversity and inclusion at the workplace: A review of research and perspectives (IIMA working papers WP2015-03-34)*. Indian Institute of Management Ahmedabad, Research and Publication Department.
O'Connor, M. Rebecca, Barrington, Wendy E., Buchanan, Diana Taibi, Bustillos, Dan, Eagen-Torkko, Meghan, Kalkbrenner, Anne, Laing, Sharon, Reding, Kerryn W., & de Castro, A. B. (2019). Short-term outcomes of a diversity, equity, and inclusion institute for nursing faculty. *Journal of Nursing Education, 58(*11), 633–640. https://doi.org/10.3928/01484834-20191021-04.
Pendry, Luise F., Driscoll, Denise M., & Field, Susannah C. (2007). Diversity training: Putting theory into practice. *Journal of Occupational and Organizational Psychology, 80*(1), 27–50. https://doi.org/10.1348/096317906X118397.
Pew Research Center. (2023). *Diversity, equity, and inclusion in the workplace*. https://www.pewresearch.org/social-trends/2023/05/17/diversity-equity-and-inclusion-in-the-workplace/.

Powell, Carmin, Yemane, Lahia, Brooks, Michelle, Johnson, Carrie, Alvarez, Al'ai, Bandstra, Belinda, Caceres, Wendy, Dierickx, Quynh, Thomas, Reena, & Blankenburg, Rebecca. (2021). Outcomes from a novel graduate medical education leadership program in advancing diversity, equity, and inclusion. *Journal of Graduate Medical Education, 13*(5), 774–784. https://doi.org/https://doi.org/10.4300/FJGME-D-21–00235.1.

Rotter, Julian B. (1954). *Social learning and clinical psychology*. Prentice-Hall.

Różalska, Aleksandra, Hauk, Mateusz, & Rudnicka, Agata. (2022). *Comprehensive gender equality and gender mainstreaming training toolbox useful for different trainee groups or national contexts*. Zenodo. https://doi.org/10.5281/zenodo.10599575.

Savolainen, Sari. (2016). Could acceptance predict commitment in organisational change? Impact of changes caused by succession from the viewpoint of non-family employees in small family firms. *Management Studies, 4*(5), 197–215. https://doi.org/10.1726 5/2328–2185/2016.05.002.

Stewart, Ian, & Joines, Vann. (2012). *TA today: A new introduction to transactional analysis*. LifeSpace Publishing Nottingham and Chapel Hill.

TAAZ Inc. (n.d.). *Diversity training isn't enough: Enhancing the effectiveness of DEI training to create a more fair and diverse workplace*. https://uwaterloo.ca/ctr-sustainability-performance-management/sites/default/files/uploads/documents/diversity-training-isnt-enough.pdf7690.

Turner, John C., Hogg, Micheal A., Oakes, Penelope J., Reicher, Stephen D., & Wetherell, Margaret S. (1987). *Rediscovering the social group: A self-categorization theory*. Basil Blackwell.

Wang, Monica L., Gomes, Alexis, Rosa, Marielis, Copeland, Phillipe, & Santana, Victor Jose. (2023). A systematic review of diversity, equity, and inclusion and antiracism training studies: Findings and future directions. *Translational Behavioral Medicine, 14*(3), 156–171. https://doi.org/10.1093/tbm/ibad061.

Zhang, Yang, & McGuire, Stephen J. J. (2021). Exploring the mechanism of diversity training through on-the-job embeddedness in a diverse workplace. *International Journal of Training Research, 20*(1), 14–25. https://doi.org/10.1080/14480220.2021.1934075.

# 11

# MAINSTREAMING INTERSECTIONAL GENDER IMPACT ASSESSMENT IN HIGHER EDUCATION INSTITUTIONS' KNOWLEDGE PRODUCTION – TRANSFORMING RESEARCH AND EDUCATION FROM WITHIN

*Mervi Heikkinen*

## Introduction and background: mainstreaming gender dimension in European higher education institutions

For the European Research Area (ERA), the main principles guiding the gender impact assessment (GIA) of studies are based on good scientific practices. These practices are implemented through the European Commission's (EC) research and innovation Framework Programmes (FPs). The FPs serve as the main research-funding instrument, with the current Horizon Europe boasting a budget of nearly 100 billion EUR. During the past 25 years, the European Commission has been formulating science policies to bridge the gap between the scientific community and society at large. In 2001, the "Science and Society" Action Plan was launched to set out a common strategy to make a better connection between science and European citizens. In 2007, the main objective of FP7's "Science in Society" initiative was to foster public engagement and a sustained two-way dialogue between science and civil society. Since 2010, the framework responding to the aspirations and ambitions of European citizens has been "Responsible Research and Innovation" (RRI).

RRI is an ambitious challenge for the creation of a Research and Innovation policy driven by the needs of society and engaging all societal actors via inclusive participatory approaches. RRI encompasses several key elements: adherence to research ethical standards, including gender equality in the scientific community and within all phases of knowledge production, assignment of responsibility to policymakers to avoid harmful effects of innovation, engaging the communities as end users of innovation, and ensuring that they have the knowledge necessary to understand the implications by furthering science education and Open Access (EC, 2012, 2020b).

DOI: 10.4324/9781003564348-18

The RRI entails that the dimensions of gender, open access, science education, ethics, and public engagement must be part of how research and innovation are introduced in all domains as well as its implications for governance. Since 2014, the projects funded under the "Science with and for Society" scheme have contributed to building cooperation between science and society, recruiting new talent for science and pairing scientific excellence with social awareness and responsibility. From 2021 onwards, Open Science became a legal obligation under Horizon Europe, accelerating the discovery process, facilitating collaboration and sharing, and thereby improving research quality and making science more impactful and central to human and societal development.

The implementation of Open Science as a central research practice requires ensuring gender equality, equal opportunities, and inclusiveness. The Coalition for Advancing Research Assessment (CoARA) and The Agreement for Reforming Research Assessment (ARRA), which facilitate open science implementation, expand this framework. They call for consideration of gender balance and broader intersectional diversity factors – including racial or ethnic origin, sexual orientation, socio-economic status, and disability – both in research teams and research content and innovation as described in The Agreement for Reforming Research Assessment (ARRA, 2022, p. 4).

Moreover, according to UNESCO (2021), Open Science has the potential to accelerate the achievement of the United Nations Sustainable Development Goals and bridge science, technology, and innovation gaps. By making the scientific process more transparent, inclusive, and democratic, it aims to fulfil the human right to science. Open Science builds on several key pillars: open scientific knowledge, open science infrastructures, science communication, open engagement of societal actors, and open dialogue with other knowledge systems. Aligned with the guiding principles of the FPs, development of tools supporting capabilities for gender-responsible research were co-designed as part of the RESET project Work Package 7, focusing on the mainstreaming of the gender dimension in the research content.

The evidence from gender-responsible knowledge production, for instance, Londa Schiebinger and her colleagues' Gendered Innovations website (Schiebinger et al., 2011–2020), demonstrates that gender and intersecting differences significantly influence all dimensions of society. The consideration of sex, gender, and intersectionality forms a fundamental part of knowledge production in higher education and research institutions-see for instance special issue on Feminist Encounters in Research and Innovation (2021, 5(2)) edited by Gabriele Griffin, Yvonne Benschop, and Liisa Husu. This is a case in global gender equality leaders in Nordic countries (Silander et al., 2021), as well as in the UK (Tzanakou et al., 2019) and in transnational Communities of Practice (Thomson et al., 2021) to mention few examples. Therefore, the analysis of gender and intersecting differences requires integration into every aspect of knowledge production (USAID, 2020; Bíziková et al., 2007) from research team composition through methodologies, project planning, research questions, and data collection, to analysis and results dissemination (Heidari et al., 2016).

### *Transforming research and education from within HEI*

The theoretical foundations of GIA draw from the feminist institutional perspective, the theory of gendered organisations, the notion of gender mainstreaming, and the analyses of gender regimes (e.g., O'Connor, 2020; Acker, 2006; Benschop & Verloo, 2015; Walby, 2020). Universities are institutions with a main function of knowledge production, wherein gender relations and gender ideologies are reconfigured and reproduced (Cullen et al., 2019). Within this environment, however, external research funding (e.g., from the EC) is one of the areas where researchers struggle to critically consider the various ways in which gender is influencing knowledge production processes. As a result, researchers are seeking help from the science support services in their universities to fill in those parts in their grant applications. In order to make this process more feasible, we developed a GIA tool, which should be considered as a "first aid": an introduction to gender dimension in knowledge production.

Scholarship on gender in knowledge production is extensive, thanks to a century-long struggle for women's full access to universities and decades-long claims for Women's and Gender Studies knowledge production on epistemic injustices, biases, and gaps. The question of mainstreaming the gender dimension in knowledge production is as deep as human consciousness. Therefore, institutional transformation processes must be deeply rooted in the very foundation of their functions – in research planning, applying for funding and conducting research – which comprise the socio-material ingredients of the production of new and ethically sound knowledge.

Furthermore, academic knowledge production should represent the interests of society in a broader manner (Bergman & Rustad, 2013) and tackle the serious problems of our times as articulated in the UN Agenda 2030 – Sustainable Development Goals (SDGs) to find solutions for them. Simultaneously, concerns regarding the quality, legitimacy, and credibility of research and Higher Education Institutions (HEIs)' overall role within society arise. Furthermore, subconscious assumptions about gender are often considered obstacles hampering fair research evaluation, thus undermining equal access, success, and retention opportunities for all. Although arguments for gender responsibility (Schiebinger et al., 2011–2020), as well as epistemic justice and epistemic contribution as a central human capability (Fricker, 2007, 2015) have recently been put forward, problems remain in their implementation in research practice.

Inspired by the work of Kimberly Crenshaw (1989), Londa Schiebinger et al. (2012–2020), and Martha Nussbaum (2011), a relatively new concept of "intersectional gender responsibility" (Heikkinen et al., 2020) has been introduced to address the acute challenges in academia related to the knowledge production required to raise awareness about sex, gender, and intersectionality dimensions in research. Intersectional gender responsibility aims to achieve three goals. The first is to promote capability, which refers to specific sensitivity through the recognition of simultaneous, intersecting sociocultural categorisations in addition to gender. The second goal is to achieve quality with regard to both scientific rigour and responsibility, which recognises and accounts for the influence of sex and gender and simultaneous sociocultural categorisations in scientific research settings and reasoning. The third objective is to achieve sustainability, which is viewed as the capacity to endure and the ability to restore and co-create important features in a knowledge society. Thus, intersectional gender responsibility contributes to rethinking gender equality together with scientific excellence, which is profoundly intertwined with issues of research ethics and integrity. Ultimately, it considers the importance of impact and implementation. In the long run, the aim is to mainstream the GIA approach to all research laboratories in the European Research Area (ERA) through GIA policy recommendations prepared in the final stage of the RESET project.

GIA is an ex-ante evaluation of the sex and gender dimensions in research activities and products, whose main objective is to determine the degree of sex and gender responsibility of a research proposal. GIA questions, analyses, sensitises, and integrates sex and gender categories into the research process wherever appropriate. In this way, GIA not only contributes to the success of an individual researcher's projects but also enhances the work of research teams, disciplines, and organisations, and advances science and society at large. Mathias Nielsen et al. (2018) present four interdependent domains, that are 1) research teams, 2) disciplines, 3) research organisations, and 4) societies at large. Furthermore, "Specific management techniques in each domain – research teams, disciplines, research organisations and societies at large – are crucial to foster the benefits associated with gender diversity in research teams, in research methods and in research questions" (Nielsen et al., 2018, p. 728). GIA is a tool available for management in each of these domains.

While the sex and gender dimensions are important factors to include, they are shaped by other sociocultural and biological factors. Therefore, in addition to sex and gender dimensions, intersectional analysis should also be considered as it may be relevant in studies involving human subjects. It also applies to people conducting research, thus giving importance to the "diversity of a research group". The way a research problem is formulated also determines which intersecting variables are required for analysis. According to Olena Hankivsky (2014), the most important categories, factors, and relationships cannot be determined a priori; rather they emerge during the research process as knowledge production takes a posteriori form. Therefore, determining the relevant intersecting identities requires sensitivity from members of the research group, which can be further improved through intersectional gender equality consultation and diversity and inclusion training. Thus, operationalisation of intersectionality sets demands for relevant expertise and opens up an opportunity to find synergies between the researchers, and the research groups, within an institution.

During the application preparation process within the Horizon Europe programme, researchers need to write three, equally important application parts – components, according to which research-funding applications are evaluated: excellence, impact, and implementation. Each component's evaluation criteria include unique aspects, ensuring that they are not assessed twice. In particular, Open Science practices as well as sex and gender dimension in research content are assessed in the excellence criterion as they are part of the scientific methodology.

Therefore, the GIA of research proposals focuses on three areas of evaluation (excellence, impact, and implementation). The evaluation process is adapted to each type of action, as specified in the European Commission's work programmes. A closer look at the evaluation criteria used in research and innovation actions (RIA) and innovation action (IA) is needed to estimate how they can impact intersectional, gender-responsible knowledge for the benefit of society.

The chapter provides a description of the GIA checklist, the ultimate goal of which is to provide a learning process on gender impact assessment, thus requiring a hands-on approach. The GIA checklist is constructively aligned to produce an in-depth understanding of what GIA is, why it is needed, and how it is carried out. This approach aims to equip researchers with competencies to use GIA as part of their normal systematic work routine. As an example, I focus on GIA in knowledge production as it was co-designed within the RESET project. The RESET project links impact and excellence criteria by enhancing the understanding of "excellence": "Our definition of scientific excellence entails supporting and sustaining inclusive, reflexive, impact-driven, societally relevant research and innovation" (Joint Statement in RESET-Project, by Paoletti et al., 2024). In other words, in order for a research project to excel, it must have both a societally relevant impact as well as effective and responsible implementation, both considered from a gender perspective.

## Mainstreaming intersectional Gender Impact Assessment (GIA) in knowledge production

Addressing the sex and gender dimensions in research and innovation entails considering these aspects throughout the entire research and innovation process – starting from the design, through execution to the implementation phase. The sex and gender dimensions bring added value to research in terms of excellence, rigour, reproducibility, creativity, and business opportunities (Nielsen et al., 2018; Schiebinger et al., 2011–2020; EU, 2011, 2014, 2020a; RESET, 2021–2024). With such an approach to science, the ultimate goal is to enhance the societal relevance of, and the drive for, responsible research and innovation (RRI). Research must contribute to building more gender-equitable and sustainable societies and, in that sense, it should set a normative imperative for all research projects to look for ways to integrate gender. This also relates to the United Nations' Agenda 2030 and the 17 identified sustainable development goals, which all require considerations to sex and gender dimension. Unfortunately, this is not the case yet. Racher Herbert and her colleagues' study (Herbert et al., 2020) in Elsevier illustrates a shortage of sex and gender dimension in SDGs-related research publications.

UN Women claims that when it comes to solving the identified sustainable development goals, sex and gender-sensitive knowledge is needed, but perhaps surprisingly they are struggling to obtain annual member states' reports to provide indicators with appropriate gender-disaggregated data (UN Women, 2018). Furthermore, EC has set gender dimension in education and research as one of the five central areas in institutional GEPs. Thus, GEPs should consist of action points improving equal access, success, and retention in HEIs, which are contributing to mainstreaming gender dimension in knowledge production generally, and consequently developing human capabilities in science for democracy building in societies and equitable futures for all (Nussbaum, 2011). Intersectionally gender-responsible knowledge production should be viewed as aligned with the GIA. In other words, the GIA approach guides researchers to develop their own capabilities to conduct research that is intersectionally gender responsible, which results in the betterment of society at large and the building of gender and diversity-responsible awareness.

The European Commission has made large investments in research and development (R&D) through the Horizon Europe scheme, which is the EU's key funding programme for research and innovation with a budget allocation of almost 100 billion euros. Under Horizon Europe, the integration of sex and gender dimensions in research and innovation undertakings is mandatory, unless it is explicitly mentioned in the topic description that it is "not a mandatory requirement". However, it is always possible to go further and include gender equality in the composition of research teams. Another step would be to incorporate gender equality (GE) and diversity training for team members and to consider these aspects as part of the strategy for disseminating research results and selecting appropriate publication channels that align with this goal while also taking into account the diversity of audiences.

Thus far, excellent guidance and the multitude of questions, indicators, and checklists that have been designed have been eye-openers for researchers, research support staff, and students (see RESET project GIA online checklist https://toolkit.wereset.eu/#/gia-checklist). For instance, the following questions may help to reflect whether a research team is diverse enough and a variety of perspectives are included in the planning of the work: Are women, men, and non-binary people introduced by full name and title during team meetings? Are both male, female, and non-binary decision-makers in the room? Are women, men, and non-binary afforded equal opportunities to speak and voice their opinions during team meetings? These general inquiries can be followed by more detailed questions about the content of the project proposal: Do the tasks defined in research involve diversity of individuals or populations as the research subject? Are the differences between men, women, and non-binary stated or defined in tasks? Gradually, through reflection, research group members learn to see, to question, and to do gender differently, that is, circulate tasks and take various responsibilities with the aim to contribute to gender equality both in knowledge production and everyday practices within a research institution.

All these mentioned different institutional gendered practices (Martin, 2003; West & Zimmerman, 1987) or ways of undoing gender or doing it differently (Butler, 2004, 2024; Deutsch, 2007) relate to various institutional policies such as the Gender Equality Plan (GEP). Thus, it is worth considering how a research project seeking EC funding would actually benefit from an implementation of GIA through an institutional GEP, which is required from all applicants and is the eligibility criterion set by EC. By this, I mean that a systematic gender equality promotion in research and innovation projects must also be included in the application, and it is necessary to align it with the institutional gender equality work (such as defined in GEPs). Furthermore, sex, gender, and intersectional dimensions must be analytically incorporated into the actual research content, while the GIA of the whole research proposal needs to be executed.

In collaboration with the RESET project's participants, we co-designed a comprehensive and innovative gender impact assessment tool – a checklist consisting of questions – which can be used to determine how an intersectional gender perspective is included in the research plan, its justification, the actual execution of the research project, and the publication of research results and their impact.

## Gender Impact Assessment of research – GIA tools

The RESET GIA approach – including GIA guidelines, checklist, and protocol – was co-designed to inform and assist HEIs in setting up a support system for local researchers to guide them in including gender dimension in research content. Additionally, a Gender Impact Assessment (GIA) online course is one of the various methods employed to expand knowledge in this regard. This is a recommended and effective way to disseminate GIA to early stage researchers. Within the RESET project, we have offered the GIA course content and syllabus to MA and PhD students as well as postdoc researchers.

The Gender Impact Assessment (GIA) approach should be applied throughout the research project as a reflective practice to ensure that intersectional sex and gender analysis is executed in each and every phase of the knowledge production as well as throughout the project execution. In this context, the following GIA actions can be identified:

1) Keyword search through literature and scholarship in the particular field.
2) Formulating research questions and objectives.
3) The analysis of knowledge gaps on gender in the research field in question.

4) Recruitment processes, with particular emphasis on female participation in decision-making and community engagement processes.
5) Gender-sensitive and diversity-sensitive data aggregation.
6) Detecting stereotypes that may introduce bias into data.
7) Elaborating on risk assessment from a gender equality point of view and designing methods to mitigate risks.
8) Gender-sensitive and diversity-sensitive reporting of research results to policymakers and practitioners, including reflection on the ways in which research advances equality.
9) Dissemination and communication activities, for example, using language that is free from bias, ensuring gender equality in panel discussions, conferences, and interviews as well as equal visibility in the media, etc.

The aim of GIA is to help researchers to foresee and elaborate on various project phases requiring their special attention already at the research design and grant application writing phases and to identify possible challenges and their remedies from gender dimension. The GIA checklist was co-designed and structured according to the European Commission's grant application structure (EC, 2021). The proposal of the GIA tool consists of three main parts devoted to excellence, implementation, and impact. Each of these three sections contains a separate set of questions to facilitate the process of research design and implementation.

### *Excellence – state of the art in research*

The excellence criterion focuses on the clarity and pertinence of the project's objectives, and the extent to which the proposed work is ambitious, innovative, and goes beyond the state of the art. There are several important issues to be taken into consideration while planning research: the soundness of the proposed methodology, the underlying concepts, models, assumptions, and interdisciplinary approaches; appropriate consideration of the sex and gender dimensions in research and innovation content as well as the quality of Open Science practices, including sharing and management of research outputs and engagement of citizens, civil society, and end users wherever appropriate. This phase also involves assessing whether the sex and gender dimensions in research and innovation content have been properly taken into account, for instance, with regard to topic choice, literature review, knowledge gaps, research question, and so on.

These 12 statements cover a broad range of research but are not exhaustive. Further considerations worthy of elaboration should be included in each research plan and application. Statements are written in an "I" form triggering personal reflection as well as responsibility: What do I really know about the topic in question? Consequently, there might be a need to revisit and elaborate some of the aspects in the GIA checklist and in the application.

**TABLE 11.1** Excellence-related set of questions

| | *Yes* | *No* | *I do not know* | *Not applicable* |
|---|---|---|---|---|
| 1. I have conducted a literature review and included sex and gender in my search keywords. | | | | |
| 2. I take into consideration the diversity of quoted authors (sex, geographical origin, etc.). | | | | |
| 3. I indicate the first name of the authors in the bibliography. | | | | |
| 4. I have considered gender implications in how I have elaborated my research question and my research goals. | | | | |
| 5. When thinking of the research or data gaps, I consider how gender may play a role in producing such gaps. | | | | |
| 6. I plan to include sex and gender-disaggregated data. | | | | |
| 7. If any differences in sex-disaggregated data exist, I ask myself whether these differences are influenced by gender roles in society. | | | | |
| 8. My way to interpret sex-disaggregated data (if they exist) includes self-reflection on the influence of stereotypes and unconscious biases. | | | | |
| 9. I consider diverse (gender, sex, age, origin, etc.) communities in the sampling. | | | | |
| 10. I have considered the gender-specific risks associated with this research and I have designed measures to mitigate against these risks. | | | | |
| 11. I have verified existing gender theories that concern the subject of my research. | | | | |
| 12. I include a gender expert/expertise in my team. | | | | |

### *Implementation – execution plan of the research*

The implementation criterion analyses the quality and effectiveness of the work plan, the foreseen risks, the appropriateness of the effort assigned to work packages, and the sufficiency of the budgeted resources overall. In particular, such crucial issues as capacity and the role of each participant to the extent to which the consortium as a whole brings together the necessary expertise are evaluated in this section. The overall objective of this phase of research is to check whether one has the sex and gender dimensions in quality and effectiveness of the work plan properly considered. Furthermore, one can reflect upon how to assemble a research team so that it is equal, diverse, and inclusive.

The 11 questions focus on working conditions that promote equal opportunities within the project and the equal outcomes of the project.

**TABLE 11.2** Implementation-related set of questions

| | *Yes* | *No* | *I do not know* | *Not applicable* |
|---|---|---|---|---|
| 1. I am collecting/using gender-disaggregated data whenever possible. | | | | |
| 2. I have designed data collection tools to take into account and challenge gender stereotypes, and social and cultural factors that may introduce gender bias into the data. | | | | |
| 3. My research team is balanced in terms of sex and diversity (nationality, age, origin, status, academic age, etc.). | | | | |
| 4. I have considered the gender balance in the project consortium or team. | | | | |
| 5. I have considered sex, gender, and diversity dimensions in the recruitment, job descriptions, and career paths of research group members. | | | | |
| 6. I will create opportunities throughout the research cycle to be reflexive and aware of my own and my team's gender assumptions, biases, and power as researchers. | | | | |
| 7. There are dimensions other than sex/gender that are important to consider. | | | | |
| 8. Even if the team is not obviously diverse (for instance, all members come from the same field, gender, ethnicity, etc.), I take into account points of view and experiences of all social groups. | | | | |
| 9. In my team, all points of view are heard and all members are listened to. | | | | |
| 10. Tasks in my team are circulated or distributed in a way that does not reproduce gender stereotypes. | | | | |
| 11. Researchers trained in gender studies are included in the research team. | | | | |

### *Impact – dissemination of the research*

The impact criterion assesses the credibility of the pathways to achieve the expected outcomes and impacts specified in the work programme, and an estimation of how significant the project's outcomes are. During this final stage of research, the suitability and quality of measures to maximise expected outcomes and impacts, as set out in the dissemination and exploitation plan, including communication activities, are critically evaluated. Consequently, at this point, we would like to assess whether the sex and gender dimensions in the pathways, impacts, and contributions have been successfully taken into account.

It is important to ask some general questions at the end of the GIA checklist about the societal effects of research and its impact on the external environment and non-academic context.

**TABLE 11.3** Impact-related set of questions

| | *Yes* | *No* | *I do not know* | *Not applicable* |
|---|---|---|---|---|
| 1. I am using appropriate terminologies and language that do not reflect gender stereotypes and do not assume only two genders. | | | | |
| 2. All research outputs will be verified for use of appropriate terminologies and language that do not reflect gender stereotypes and that do not assume only two genders. | | | | |
| 3. When considering authors, inviting keynotes, planning publications, and providing visibility for researchers and their work on websites, I pay attention to gender balance. | | | | |
| 4. I valorise all members of the research team in the dissemination phase (authors, publications, website, keynote, etc.). | | | | |
| 5. The sex/gender dimension is included in the presentation of findings. | | | | |
| 6. Research reports/publications/outputs will be revised by a gender expert. | | | | |
| 7. I have included gender equality training for the project staff. | | | | |
| 8. I consider that the results of my research (project) can have different effects on men and women, boys or girls. | | | | |
| 9. My research can contribute to the advancement of gender equality in society. | | | | |

## Discussion and critical reflection: gender responsibility of knowledge as an epistemological and ethical challenge

Each and every GIA exercise offers a unique opportunity to find various hidden meanings for sex and gender as well as for intersecting social categories. Based on our experiences within the RESET project, exploration of an intersectional gender impact is essential for the production of new knowledge. Innovative ways to apply sex and gender perspectives and intersectional analyses into research should be constantly developed and improved. The purpose of applying GIA is not just about curiosity but also to enhance the quality of knowledge production and, finally, our understanding of the world and its various phenomena in a more detailed and accurate way.

GIA has a clear message about the importance of including the sex, gender, and intersectionality dimensions thoroughly in the knowledge production process. This goes in line with the criticism of scientific objectivity and the requirement for "strong objectivity" presented by Sandra Harding in 1986 in her renowned book *The Science Question in Feminism* (Harding, 1986, 2004). According to Harding, researchers are always situated and therefore they need to both explicate their positionality as well as consider those in the margin, for instance, in the data collection. Sex, gender, and intersectional dimensions are significant considerations throughout the entire knowledge production process – starting from its design of excellence, implementation, and impact. We simply cannot afford to make bad science, as Londa Schiebinger puts it (2011–2020), by producing biased results that may be life-threatening for half of the population. Instead, we need to incorporate sex, gender, and intersectional dimensions into our research and acknowledge them as central to accurate knowledge production. Therefore, response-ability (Barad, 2000; Haraway, 2016; Higgins, 2021) – an ability to respond with care to collective needs of knowledge, essential for informed actions – and intersectional gender responsibility (Heikkinen et al., 2020) should be at the core of ethically sound knowledge production and higher education for sustainable futures.

Mainstreaming gender dimension in research content is a real challenge due to a lack of experts with sufficient training in the areas of interdisciplinary women's and gender studies and feminist epistemologies. Furthermore, it is important to note that at present not all HEIs within the European Research Area have a Chair in Gender Studies or even offer any courses on the subject for students or staff members. As a consequence, appropriate scholarship is lacking from various HEIs resulting in a hindrance for considering gendered innovations, or gender equality as central in scientific excellence, and overall, ethically gender response-ability and sustainability in science.

The GIA approach – co-designed, implemented, and further developed throughout the RESET project – was originally intended and designed to be used in a consultation session together with gender studies scholars from various fields or trained experts from the research support services. However, due to the shortage of gender experts within European HEIs, a research group or a lab may also implement GIA independently, just by using the GIA checklist as a heuristic tool to gain new views by assessing and explicating the gender dimensions in their planned research project. The utmost intention of GIA is to advance science and society at large through improved academic knowledge and knowledge production practices, but it also contributes to the success of research teams' proposals, gender response-able research projects and ethically sound HEI practices, as well as individual researchers' competences to conduct scientific research with refined excellence. In addition to providing support to strengthening capabilities for gender-responsible research, we hope that GIA tools are fostering the scientific curiosity of scholars to analyse and interpret gender and intersecting differences in many disciplines and new research areas.

## References


Acker, Joan. (2006). Inequality regimes: Gender, class, and race in organizations. *Gender & Society*, *20*(4), 441–464. https://doi.org/10.1177/0891!I243206289499

ARRA. (2022). *Agreement for reforming research assessment.* https://coara.eu/app/uploads/2022/09/2022_07_19_rra_agreement_final.pdf

Barad, Karen. (2000). Reconceiving scientific literacy as agential literacy, or learning how to intra-act responsibly within the world. In Sharon Traweek & Roddey Reid (Eds.), *Doing science + culture*, 221–258. New York: Routledge.

Benschop, Yvonne, & Verloo, Mieke. (2015). Feminist organization theories. Islands of treasure. In Greenwood Michelle, Mir Raza, & Wilmott Hugh (Eds.), *Routledge companion to philosophy in organization studies*, 100–113. New York: Routledge.

Bergman, Solveig, & Rustad, Linda M. (2013). *The Nordic region – a step closer to gender balance in research? Joint Nordic strategies and measures to promote gender balance among researchers in academia.* Nordic Council of Ministers. https://norden.diva-portal.org/smash/get/diva2:702842/FULLTEXT01.pdf

Bíziková, Lívia, Sedová, Tatiana, & Szapuová, Mariana. (2007). *Why gendered science matters. How to include gender dimension into research projects.* http://www.klemens.sav.sk/fiusav/cecwys/docs/why_gendered_science_matters.pdf

Butler, Judith. (2004). *Undoing gender*. Routledge.

Butler, Judith. (2024). *Who's afraid of gender*. Allen Lane.

Crenshaw, Kimberly W. (1989). *Demarginalizing the intersection of race and sex: A black feminist critique of antidiscrimination doctrine. Feminist theory and antiracist politics*, 138–167. University of Chicago Legal Forum.

Cullen, Pauline, Marx Ferree, Myra, & Verloo, Mieke. (2019). Introduction to special issue: Gender, knowledge production and knowledge work. *Gender, Work and Organisations*, *26*, 765–771. https://onlinelibrary.wiley.com/doi/10.1111/gwao.12329

Deutsch, Francine M. (2007). Undoing gender. *Gender & Society*, *21*(1), 106–127. https://doi.org/10.1177/0891243206293577

European Commission: Directorate-General for Research and Innovation. (2011). *Toolkit gender in EU-funded research. Yellow window management consultants, engender, genderatwork.* Luxembourg: Publications Office of the European Union. https://www.ki.si/fileadmin/user_upload/KINA24840ENC_002.pdf

European Commission: Directorate-General for Research and Innovation. (2012). *Responsible research and innovation – Europe's ability to respond to societal challenges*. Publications Office, https://data.europa.eu/doi/10.2777/11739

European Commission: Directorate-General for Research and Innovation. (2014). *Gender in EU-funded research: Toolkit.* Luxembourg: Publications Office. https://data.europa.eu/doi/10.2777/74150

European Commission: Directorate-General for Research and Innovation. (2020a). *Gendered innovations 2 – how inclusive analysis contributes to research and innovation – policy review*. Publications Office of the European Union. https://data.europa.eu/doi/10.2777/316197

European Commission, Directorate-General for Research and Innovation. (2021). *Horizon Europe proposal evaluation standard briefing version 4.0 26.09.2022*. https://ec.europa.eu/info/funding-tenders/opportunities/docs/2021-2027/experts/standard-briefing-slides-for-experts_he_en.pdf

European Commission: Directorate-General for Research and Innovation, Delaney, N., Iagher, R., & Tornasi, Z. (2020b). *Institutional changes towards responsible research and innovation – achievements in Horizon 2020 and recommendations on the way forward*. Publications Office, https://data.europa.eu/doi/10.2777/682661

Fricker, M. (2007). *Epistemic injustice: Power and the ethics of knowing*. Oxford Scholarship Online, https://doi.org/10.1093/acprof:oso/9780198237907.001.0001

Fricker, M. (2015). Epistemic contribution as a central human capability. In *The equal society: Essays on equality in theory and practice*, 73–90. Lanham: Lexington Books.

Griffin, Gabriele, Benschop, Yvonne, & Husu, Liisa. (2021). Feminist encounters: A journal of critical studies in culture and politics. *Special Issue on Feminist Encounters in Research and Innovation, 5*(2). https://www.lectitopublishing.nl/feminist-encounters

Hankivsky, Olena. (2014). *Intersectionality 101*. Vancouver, Canada: Institute for Intersectionality Research & Policy, Simon Fraser University. https://bccampus.ca/wp-content/uploads/2020/07/Hankivsky-Intersectionality101–2014.pdf

Haraway, Donna. (2016). *Staying with the trouble: Making Kin in the Chthulucene*. Durham, NC: Duke University Press.

Harding, Sandra. (1986). *The science question in feminism*. Cornell University Press.

Harding, Sandra. (2004). Rethinking standpoint epistemology: What is "strong objectivity"? In Sandra Harding (Ed.), *Feminist standpoint theory reader. Intellectual and political controversies*, 127–140. Routledge.

Heidari, Shirin, Babor, Thomas F., De Castro, Paola, Tort, Sera, & Curno, Mirjam. (2016). Sex and gender equity in research: Rationale for the SAGER guidelines and recommended use. *Research Integrity and Peer Review, 1*, 1–9. https://doi.org/10.1186/s41073-016-0007-6

Heikkinen, Mervi, Pihkala, Suvi, Pääsky, Leena, & Harmoinen, Sari. (2020). *Intersectional gender-responsibility in STEM: Co-creating sustainable Arctic knowledge production*, 175–188. Arctic Yearbook. https://arcticyearbook.com/images/yearbook/2020/Scholarly-Papers/9_Heikkinen_et_al.pdf

Herbert, Rachel, Falk-Krzesinski, Holly, & Plume, Andrew. (2020). Sustainability through a gender lens: The extent to which research on UN sustainable development goals (SDGs) includes sex and gender consideration. http://dx.doi.org/10.2139/ssrn.3689205.

Higgins, Marc. (2021). Response-ability revisited: Towards re (con) figuring scientific literacy. In Unsettling responsibility in science education: Indigenous science, deconstruction, and the multicultural science education debate, 271–316. https://doi.org/10.1007/978-3-030-61299-3_7

Martin, Patricia Yancey. (2003). "Said and done" versus "saying and doing": Gendering practices, practicing gender at work. *Gender & Society, 17*(3), 342–366. https://doi.org/10.1177/0891243203017003002

Nielsen, Mathias Wullum, Bloch, Carter Walter, & Schiebinger, Londa. (2018). Making gender diversity work for scientific discovery and innovation. *Nature Human Behaviour, 2*, 726–734. https://doi.org/10.1038/s41562-018-0433-1.

Nussbaum, Martha C. (2011). *Creating capabilities. The human development approach*. Harvard University Press.

O'Connor, Pat. (2020). Why is it so difficult to reduce gender inequality in male-dominated higher educational organizations? A feminist institutional perspective. *Interdisciplinary Science Reviews, 45*(2), 207–228.

Paoletti, Marion, Junca, Ninon, Radchuk, Maryna, Bergstedt, Friederike, Niebel, Viktoria, Karachaliou, Elena, Argyropoulou, Laura, Różalska, Aleksandra, Heikkinen, Mervi, Matias, Marisa, Magalhães, Sara Isabel, & Forest, Maxime. (2024). *Joint statement of top management on their engagement for equality, diversity and excellence in research.* Zenodo. https://doi.org/10.5281/zenodo.14646461.

RESET. (2021–2024). *Redesigning equality and scientific excellence together-project web-site*. https://wereset.eu/

RESET-Project. (2024). *Gender impact assessment on-line checklist.* https://toolkit.wereset.eu/#/gia-checklist

Schiebinger, Londa, Klinge, Ineke, Paik, Hee Young, Sánchez de Madariaga, Inés, Schraudner, Martina, & Stefanick, Marcia. (Eds.). (2011–2020). *Gendered innovations in science, health & medicine, engineering, and environment.* http://genderedinnovations.stanford.edu/index.html

Silander, Charlotte, Haake, Ulrika, Lindberg, Leif, & Riis, Ulla. (2021). Nordic research on gender equality in academic careers: A literature review. *European Journal of Higher Education*, *12*(1), 72–97. https://doi.org/10.1080/21568235.2021.1895858

Thomson, Aleksandra, Palmén, Rachel, Reidl, Sybille, Barnard, Sarah, Beranek, Sarah, Dainty, Andrew R. J., & Hassan, Tarek M. (2021). Fostering collaborative approaches to gender equality interventions in higher education and research: The case of transnational and multi-institutional communities of practice. *Journal of Gender Studies*, *31*(1), 36–54. https://doi.org/10.1080/09589236.2021.1935804

Tzanakou, Charikleia, & Pearce, Ruth. (2019). Moderate feminism within or against the neoliberal university? The example of Athena SWAN. *Gender Work & Organization*, *26*(8), 1–21. https://doi.org/10.1111/gwao.12336

UN Women, the United Nations Entity for Gender Equality and the Empowerment of Women. (2018). *Turning promises into action: Gender equality in the 2030 agenda for sustainable development.* https://www.unwomen.org/en/digital-library/publications/2018/2/gender-equality-in-the-2030-agenda-for-sustainable-development-2018

UNESCO. The United Nations Educational, Scientific, and Cultural Organisation. (2021). *Recommendation on open science*. https://doi.org/10.54677/MNMH8546

USAID. The United States Agency for International Development. (2020). *Gender and social inclusion checklist: What do I need to consider?* https://www.usaid.gov/sites/default/files/documents/Power-Africa-Gender-Social-Inclusion-Checklist.pdf

Walby, Sylvia. (2020). Varieties of gender regimes. *Social Politics: International Studies in Gender, State and Society*, *27*(3), 414–431. https://doi.org/10.1093/sp/jxaa018

West, Candace, & Zimmerman, Don H. (1987). Doing gender. *Gender & Society*, *1*(2), 125–151. https://doi.org/10.1177/0891243287001002002

# 12

# THE CONSTELLATION OF COPS AS GENDER CHANGE AGENTS

## The meaning of communities of practitioners now and in the future

*Tuija Lämsä*

### Introduction

Organisational learning and knowledge management are crucial when inventing new practices and integrating an equality culture through GEPs (Gender Equality Plans). In this chapter, I will examine how learning takes place at different levels (individual level, team/CoP level, and organisation/university level) and how competencies develop in relation to equality issues. Establishing databases and digital tools to support gender equality policymaking, as well as implementing new practices and policies in partner organisations, requires in-depth research on knowledge management systems.

When identifying Communities of Practice (CoPs) within each university, it is important to recognise that, alongside formal CoPs (such as Gender Equality Boards) dedicated to institutional gender equality work, there are also opportunities to discover or create more informal CoPs. These less formal groups provide invaluable insights, knowledge, and expertise – both explicit and tacit – on equality and gender-related issues.

A Community of Practice (CoP) on gender equality aims to enhance scientific excellence in policymaking by creating sustainable operational models and processes that ensure the continuity of Gender Equality Plan (GEP) initiatives beyond the project's completion. These action plans aim to drive structural and cultural change. Additionally, CoPs create opportunities to unite various parties – individuals, researchers, students, and stakeholders – who are passionate about the topic of gender and equality issues and who can help promote gender equality in academic careers and contribute to designing a diverse and inclusive work environment.

DOI: 10.4324/9781003564348-19

## Defining Communities of Practice

CoPs can be defined as informal social structures with long-standing traditions, from ancient tribes to today's informal groups. According to Wenger-Trayner Etienne and Wenger-Trayner Beverly (2015), Communities of Practice are groups engaged in collective learning within a shared area of interest or activity. Examples include a tribe working together to sustain their survival, a collective of artists experimenting with new forms of expression, engineers collaborating on similar challenges, a group of students shaping their identities, a network of surgeons innovating surgical techniques, or a cohort of first-time managers supporting one another as they navigate new responsibilities.

### *The characteristics of the Communities of Practice*

Not everything called a community is a community of practice. Three characteristics are crucial to define a CoP (Wenger-Trayner & Wenger-Trayner, 2015):

1) **The domain**: a community of practice is more than just a social group or a network of connections; it is characterised by a shared domain of interest that gives it a distinct identity. Membership involves a commitment to this domain and a joint competence that sets its members apart from others. While the domain may not always be recognised as "expertise" by those outside the community, members deeply value their collective knowledge and learn from one another, even if their expertise is unnoticed or underappreciated by others.
2) **The community**: members of a community of practice actively engage in their shared domain through collaborative activities, discussions, mutual support, and information exchange. These interactions foster relationships that enable learning from one another. Simply sharing a job title or role does not constitute a CoP; it is the meaningful interaction and collective learning that define it.
3) **The practice**: a community of practice is distinct from a community of interest, such as, for example, a group of people who enjoy the same films. CoP members are practitioners who actively engage in their domain, developing a shared repertoire of resources that includes experiences, stories, tools, and methods for addressing recurring challenges – essentially, a collective practice. Building this common practice requires time and ongoing interaction.

The combination of these three core elements – a domain, a community, and a practice – defines CoPs. While all CoPs share these foundational elements, they can vary widely in form. Some are small and intimate, while others are large, with a central core group and many peripheral members. They may be local or span global boundaries. Interaction can occur primarily face-to-face, predominantly online, or through a mix of both. Some CoPs operate within a single organisation, while others bring together members from multiple institutions. In addition, some are formally recognised and supported with resources, while others remain entirely informal or even invisible. A good example of such formalised CoPs is RESET's GEBs, which are officially part of the organisation's structure and naturally also resourced so that their operation is continuous even in the longer term.

In the higher education context, these key characteristics of CoPs (domain, community, and practice) can be analysed and verified in several ways. The domain represents the shared purpose or interest that brings members of a CoP together. In HEIs, this could encompass a variety of focuses, such as teaching excellence, research on a specific topic, administrative efficiency, or the promotion of diversity and inclusion, as was the case with RESET's Gender Equality Boards. Different challenges can also be faced. Multiple positions and hierarchies mean that faculty, managers, and administrative staff have diversified job roles and competing priorities, which may fragment their focus on the domain. Also, not all members may be equally committed to common goals, especially in such areas as equality and diversity, which may not align with everyone's priorities or values.

When defining the university community, consideration is given to fostering relationships, cooperation, and a sense of belonging among members despite their multiple roles. It is important to build trust across different responsibilities and hierarchies, given the power dynamics between tenured faculty, junior staff, and administration that can inhibit open dialogue. Members of different scientific fields or research positions may also struggle to find a common ground for their activities. In the university context, practice refers to the shared knowledge, tools, and processes that members of the CoP develop and refine over time. However, they can face challenges such as disciplinary silos, as departments or laboratories often operate in isolation, making it difficult to share best practices across fields. In addition, some members may lack the time, motivation, or skills to contribute meaningfully to CoP activities. In RESET, GEBs faced these challenges since they were quite varied in their composition. These different standings lead to rich interactions but also to some tensions regarding the priorities that should be envisaged by such a community.

HEIs can adapt the CoP framework by addressing hierarchical challenges, encouraging broad-based engagement, and ensuring inclusivity. This involves balancing flexibility with structure and creating opportunities for collaboration across diverse roles and interests.

### *The dimensions of the Communities of Practice*

There are also several "parameters" or "dimensions" to describe the functions of different kinds of CoPs. Many communities perform all these aspects equally, but a closer look at how the community is structured and operates reveals that it tends to focus mainly on one or two functions. Table 12.1 illustrates the dimensions through which CoPs are described.

**TABLE 12.1** The dimensions of the Communities of Practice (Lämsä, 2008)

| *Dimensions* | *Description* |
|---|---|
| Natural vs. contrived | Emergent vs. management initiated; the stage of the formality |
| Interest | Personal and professional interests |
| Motivation | Intrinsic vs. extrinsic |
| Intensity of contact | Frequency and duration of meetings/communication |
| Voluntariness of participation | The amount of mandatory participation, the stage of participation |
| Status and power | Core or marginal status, community's power and influence |
| Know-how | Communities of Practice specific knowledge and know-how |

According to Wenger (1998), members of a community are informally connected through their shared activities – ranging from casual discussions to collaboratively solving complex problems – and the learning that arises from these interactions. Unlike a community of interest or a geographical community, a CoP is characterised by a shared practice. It is defined along three key dimensions. First, the purpose is the shared domain of interest understood as a joint enterprise and is continuously shaped and renegotiated by its members. Second, the functioning refers to a mutual engagement binding members together and creating a cohesive social entity. And, third, the output – a shared repertoire of resources developed over time, including routines, tools, vocabulary, sensibilities, and styles, which represent its collective capability.

CoPs emerge around issues that resonate with their members and with their practices shaped by members' shared understanding of what matters most (Wenger, 1998). While external constraints or directives may influence this understanding, the practices that develop are the community's unique response to these influences. Even when a CoP aligns its actions with external directives, it is the community itself – not the directive – that defines and shapes the practice. This makes CoPs inherently self-organising systems.

### *Elements of Communities of Practice*

Communities can be understood as comprising three key elements: people, places, and things (Lesser et al., 2000). People are the foundation of any community, as they engage with one another around shared interests, issues, or needs. Moreover, communities require a gathering space for interaction. Traditionally, this has been a physical location where members exchange ideas and insights. However, with the advent of the digital age, these spaces can also be virtual, enabling effective interaction across time and geographical boundaries. Wenger (2000) highlights the importance of "critical activity" within communities, which involves managing "boundary objects" or shared resources. In the context of a workplace, these boundary objects include rules, norms, procedures, tools, and artefacts that facilitate task completion. They play a vital role in documenting and sharing knowledge and guiding how work is performed.

### *The voice of Communities of Practice*

Participation in communities is often voluntary, with individuals playing various roles. CoPs are based on informal relationships between members that emerge as individuals work together to solve problems, although they might be influenced by their formal roles and relationships. People committed to these groups share common goals and interests, using the same tools and language to engage in joint practices. Fluid CoPs (Brown & Duguid, 1991) allow organisational members to hold a legitimate peripheral status, where they can observe and develop their skills. As Dougherty (1995, p. 115) notes, "competencies do not exist apart from the people" who develop them, "nor from the social processes of interpretation and construction through which people make their experiences meaningful" (1995, p. 115). Organisational theorists Prahalad and Hamel (1990) suggest that an organisation is defined by its "core competencies". Therefore, the constellations of CoPs that embody these competencies are what give organisations their identity in terms of organisational know-how (Snyder, 1996).

## Communities of Practice in organisations

CoPs can be found within any organisation. According to Wenger (1998), since membership is based on active participation rather than formal roles, these communities are not confined by organisational affiliations and can transcend institutional structures and hierarchies. CoPs are not a new type of organisational unit; instead, they represent a different perspective on an organisation's structure, focusing on shared learning and knowledge-building among members rather than specific projects or social connections. What sets CoPs apart from other groups in organisations is how they define their purpose, maintain continuity over time, and establish their boundaries.

A CoP differs from a business or institutional unit in that it defines itself through action. Members collectively shape their understanding of the practice, resulting in a more dynamic and complex definition than a formal institutional unit. Consequently, the boundaries of a CoP are more flexible than those of an organisational unit. Membership includes anyone who participates in and contributes to the practice, with varying levels of involvement.

Unlike a team, a CoP is united by shared learning and interest rather than specific tasks. It is centred on knowledge rather than assignments and exists because its members find value in participating. The life cycle of a CoP is determined by the benefit it offers to its members, not by institutional timelines. A CoP does not simply form at the start of a project or dissolve when the task ends – it evolves over time and may persist long after a project or team has concluded.

Individuals belong to CoPs while also being part of other organisational structures. In an organisation's formal institutional units, they contribute to the organisation's structure. In teams, they manage projects. In networks, they build relationships. In CoPs, they develop knowledge and expertise that enhance their effectiveness in the other roles they perform. This informal web of communities and shared practices is what underpins the functionality and success of the formal organisation.

### *The meaning of Communities of Practice*

CoPs play a vital role in any organisation, but they are especially critical for institutions that view knowledge as a core asset, such as HEIs. While CoPs are essential to the operation of any organisation, they become particularly pivotal in those that view knowledge as a fundamental resource. In this context, a successful institution consists of a network of interconnected CoPs, each focusing on specific areas of expertise and competence – ranging from promoting equitable treatment for all members (students and staff alike) to ensuring high-quality research output and fostering national and international societal impact.

As highlighted by Lämsä and Nätti (2011), CoPs play a crucial role in knowledge-intensive organisations by promoting knowledge-sharing and fostering innovation, making them integral to knowledge creation and diffusion in such contexts. CoPs serve several key functions in the creation, retention, and dissemination of knowledge within an organisation, they facilitate the exchange and interpretation of information, preserve institutional knowledge, nurture essential competencies, and provide a sense of identity for their members (Wenger, 1998):

- *CoPs play a crucial role in facilitating the exchange and interpretation of information within knowledge-intensive environments.* Members of these communities share a common understanding and knowledge base, which helps define their focus and strengthens group identity. However, creating shared knowledge and meanings can sometimes hinder cooperation between different CoPs, as the embedded practices within a community may limit cross-group interaction (Wenger, 1998, p. 3).
- *CoPs also function as vital knowledge retainers, especially in organisations dealing with both tacit and explicit knowledge.* The nature of the knowledge being shared is essential in shaping how effectively it is transferred, accumulated, and retained within an organisation, as well as how widely it disseminates across various areas. As noted by Davenport et al. (1998), knowledge extends beyond mere information; it encompasses experience, context, interpretation, and reflection. This elevates it to a high-value form of information, primed for application in decision-making and action (Wenger, 1998, p. 3).

- *CoPs act as stewards of competence,* keeping organisations at the forefront of development by fostering innovation, sharing ideas, and addressing problems collaboratively. Members of CoPs invest their professional identities in these communities, making them dynamic, forward-thinking entities. Unlike temporary teams or business units, CoPs provide a stable environment where identities are shaped, which is crucial for learning and creativity. Supporting the development of these identities is the key for organisations to harness the full potential of their employees (Wenger, 1998, p. 3).

The above-mentioned features, functions, and competencies are needed in all kinds of CoPs created during the RESET project. Good examples of these are, for instance, Gender Equality Boards (GEBs) and Gender Equality Committees, which were set up within RESET universities. In these communities, members share a common understanding and knowledge base (for example, of gender equality issues), which helps define their focus and strengthens group identity. These CoPs also function as vital knowledge retainers and as stewards of competence, sharing ideas and addressing problems collaboratively, and spreading awareness and disseminating new knowledge created in the groups to the entire organisation, for example, through other committees, meetings, and events, at the department, unit, faculty, and organisational level. CoPs can be seen to serve as the natural social framework for the ownership and exchange of knowledge and expertise (Lämsä, 2014).

Different members of an organisation can take action in their own domains to support CoPs and to maximise the benefits they can provide (Wenger, 1998). For example, deans, deans for education, and heads of departments should make sure that people are able to participate in the right CoPs so they can sustain the expertise they need to contribute to projects and teaching activities. Knowledge managers (professionals responsible for overseeing the effective creation, sharing, and utilisation of knowledge within an organisation) ought to go beyond creating informational repositories that take knowledge from being a neutral, inactive commodity, towards supporting the whole social and technical ecosystem in which knowledge is retained and created. Strategists should establish two-way connections between CoPs and organisational strategies, leveraging change managers to ensure successful and sustainable transitions. These professionals drive new strategies, processes, and cultural shifts, helping to build practices and communities that create meaningful, lasting change.

## The role of the organisation and the managerial implications in relation to Communities of Practice

In discussions about CoPs, it is often emphasised that they are natural and cannot be managed like a project or team. While I agree that CoPs are largely organic, they can still be supported by the organisation. They can function like volunteer communities that continually redefine themselves in a natural manner, but they can also be formalised groups integrated into the organisation's structure and operations. However, those who support these communities must understand the conditions that encourage their development and create an environment where they can thrive.

According to Moran and Brightman, change management has been defined as "the process of continually renewing an organization's direction, structure, and capabilities to serve the ever-changing needs of external and internal customers" (2001, p. 111). Change management helps ensure that CoPs not only function effectively but also drive meaningful and sustainable organisational development in HEIs. It also refers to a structured approach for guiding individuals, teams, and organisations through transitions to achieve desired outcomes. It involves planning, implementing, and monitoring strategies to help people adapt to changes in processes, systems, structures, or culture. The goal of change management is to minimise resistance, reduce disruptions, and ensure that the change is effective, sustainable, and aligned with organisational objectives. It encompasses both the technical aspects of change and the human element, focusing on communication, training, and stakeholder engagement (see the chapter by Rudnicka-Reichel and Różalska in this book). However, this process is much more complicated as getting employees to change their behaviour is often a major challenge and can sometimes feel almost impossible. People are not only central to successful organisational change, they also present the greatest challenge in mastering it. Employees have diverse backgrounds, skills, abilities, personalities, dispositions, challenges, motivations, and aspirations, all of which influence their actions. Yet, when planning change, leaders often fail to account for these differences.

At the community level, CoPs help establish a common language, methods, and procedures around specific knowledge and competencies – people in Communities of Practice manage change more effectively as part of a collective than alone, individually. The knowledge and expertise of the community are distributed among many employees, providing a significant benefit: the retention of knowledge when some employees leave the organisation. CoPs preserve the tacit aspects of knowledge that formal systems, such as databases or procedures, cannot capture, making them ideal for onboarding newcomers.

Even when communities standardise their working processes, they can adapt to local circumstances, thus remaining useful to practitioners. They also enhance access to expertise across organisational levels, serving as hubs for the exchange and interpretation of information and knowledge. Because their members share a common understanding, they are aware of what is relevant to communicate and how to present information effectively. As a result, a community of practice that spans the organisation becomes an ideal channel for disseminating information, such as best practices, tips, and feedback, across organisational boundaries.

## Communities of Practice in RESET universities during and after the project

In the RESET project, it was important for partner organisations to map practitioners and experts interested in and engaged with gender equality and diversity issues at their institutions, in research groups, at faculties, in laboratories, etc. For example, identifying experts committed through their position, research, or other interests was valuable, as these individuals can be recognised as "gender change agents". These goals were well achieved during the project by the partner institutions. The members of CoPs have played a key role in exchanging ideas, sharing interests and experiences, building knowledge, understanding, and awareness of gender equality and diversity. In the RESET project, CoPs have been either a group of professors working on similar problems (e.g., gender equality) or a group of doctoral students or researchers (e.g., interested in issues of intersectionality), which can be very informal in nature.

CoPs can also be more formal groups at the organisational or network level. During the RESET project, various CoPs were created and developed. Each partner institution established GEB (Gender Equality Board) or GEC (Gender Equality Committee). In addition, some partners created a CoP on GIA (Gender Impact Assessment) that supported the focus on gender integration into research. Other partners organised CoPs around particular issues, such as work–life balance (WLB), scientific excellence, Gender Equality Plan, anti-discrimination mechanisms, training on GE and diversity, teaching gender studies, and so on. Each consortium member had their own RESET team that can also be considered a CoP. Naturally, our own RESET project CoP consists of all the partner universities that participated in the project.

RESET collaborated with numerous EU-funded projects dedicated to advancing gender equality in research and academic environments. CoPs play a crucial role in these efforts and are actively utilised in various sister projects. For instance, the international EU-funded ACT on Gender project (Promoting Communities of Practice to advance knowledge, collaborative learning, and institutional change on gender equality in the European Research Area) established and supported eight CoPs[1] to drive gender equality initiatives within universities, research centres, and funding organisations across the European Research Area. Similarly, the INSPIRE project recognised CoPs as effective tools to combat the isolation of practitioners, coordinate equality efforts across organisational boundaries, and build capacity for transformative change.[2]

Another example is GENDERsafe, which promotes comprehensive policies to address gender-based violence in research and higher education. GENDERsafe has created a CoP open to a diverse group of stakeholders, including equality, diversity, and inclusion officers; human resources and administrative staff; institutional leaders; national government representatives; trade unions and gender equality organisations; researchers; professors; students; and university activists. The GENDERsafe CoP facilitates exchange and mutual learning on institutional policies addressing gender-based violence. It aims to strategically tackle gaps in institutional responses through a trauma-centred approach, addressing the continuum of violence. Its overarching goal is to promote EU-wide policy coordination by contributing to the development of a model policy framework, including a code of conduct that fosters a zero-tolerance approach to gender-based violence in academia and research.[3]

### *The future of CoPs after the RESET project*

The future of CoPs after the project relies on several factors, such as how effectively CoPs are sustained and integrated into the organisation's culture and structure. It also depends on whether the organisation's management is committed to fostering lasting change and dedicated to maintaining and further developing CoPs. Here are some possible outcomes and considerations for their continuation:

*Institutionalisation of CoPs.* The institutionalisation of CoPs involves creating a sustainable plan or strategy to institutionalise these CoPs. This includes embedding best practices created and innovated within CoPs during the project into the broader organisational structure. If organisations recognise the value CoPs bring in terms of knowledge-sharing, innovation, and problem-solving, they may formally integrate them into the institutional framework. CoPs can evolve from being project-specific to becoming permanent structures that transcend individual projects, contributing to ongoing learning and improvement.

*Ongoing knowledge transfer.* Even after a project ends, CoPs can play a vital role in retaining and transferring knowledge within the organisation. As employees move on to new projects or roles, CoPs can help ensure that valuable insights, lessons learned, and best practices are shared, preventing knowledge loss.

*Self-sustaining communities.* CoPs can become self-sustaining without formal organisational support. When members derive value from their ongoing collaboration, they are likely to continue to meet, exchange ideas, and support each other's professional growth. These organic CoPs can thrive as informal networks that foster continuous learning. However, support for CoPs is necessary, especially in the long term, to sustain their development. Long-term commitment is important for creating group identity, enhancing communication, and facilitating the restructuring of partner organisations. The goal is to scale up CoPs for systematic change in areas such as careers, decision-making, and the integration of the gender dimension.

*Expansion across projects.* Successful CoPs may expand their reach beyond the original project, influencing future initiatives. Members can share their experiences with other teams, and the practices developed within the CoP can be adapted more broadly across the organisation.

*Challenges in sustainability.* Without proper support, CoPs can dissolve after a project concludes. Challenges such as lack of time, resources, or organisational commitment can make it difficult to maintain momentum. To ensure the future of CoPs, leaders must provide ongoing support, such as resources, platforms for collaboration, and recognition of the value that CoPs bring.

One important thing we learned during the RESET project was that we had a great opportunity for interdisciplinary cooperation. Since RESET involved seven large multidisciplinary universities from different parts of Europe and the members of the project groups also represented several different research fields (e.g., social sciences, humanities, IT sciences, psychology, and management and organisational studies), we observed differences in structures, functions, and processes. However, these were not obstacles. On the contrary, since the common goal of the project was to meet the challenges for gender equality in HEIs (such as creating a gender- and diversity-friendly environment, monitoring and evaluating changes towards gender-sensitive organisations), this interdisciplinary collaboration was opening new perspectives and strategies, as well as best practices that can be implemented in organisational operations and processes in future.

## Discussion

Developing knowledge and fostering a deeper understanding of gender equality issues are essential for achieving excellence in research projects and building core competencies at both local and national levels. This process begins by acknowledging the unique identity of each partner organisation, its operational culture, and the societal context at the national level. In addition, sharing best practices at institutional and national levels in the areas of gender equality and decision-making is vital. As members of CoPs simultaneously engage with other organisational structures, they play a key role in disseminating and transferring their experiences and knowledge within their organisations, thereby amplifying the impact of CoPs.

In the context of higher education, the key characteristics of CoPs (domain, community, and practice) can be identified and assessed through various means. HEIs should think through their strategies: for example, they need to clearly articulate the domain's relevance to all stakeholders, emphasising shared benefits (e.g., better learning outcomes or improved institutional reputation) and allowing for flexibility in how individuals engage with the domain to accommodate diversified interests and levels of commitment. Universities could also create neutral spaces for interaction (e.g., cross-departmental workshops and online platforms) and implement various kinds of training sessions where hierarchical barriers are minimised. One of the main goals of these training sessions is to build the capacity of institutional stakeholders, who are the main drivers of change, and thus their confidence and legitimacy. The local training sessions can also be open to the public in the form of, for instance, a webinar on sexual violence or a public event on gender-based violence in academia. It is also important to utilise technologies (e.g., creating collaborative platforms) to document and widely share practices across the institution.

By recognising and rewarding the contributions made to the community, either through official recognition, funding, or career incentives, it increases the continuity of operations and helps to develop the entire community.

HEIs often comprise individuals with varying levels of interest or investment in issues of equality and diversity. While some may be active advocates, others may show indifference or resistance. Strategies for inclusion could be, for example, awareness-raising activities and training or regular workshops to expand knowledge and skills about the importance of DEI culture in and through CoPs.

Ongoing support for CoPs is essential at various stages, especially in the long run. This long-term commitment is key to building group identity and successful communication and to supporting structural change within partner organisations. Scaling up CoPs for systematic change is crucial for careers, decision-making, and gender integration. RESET universities strive to systematically include gender aspects into policies, programmes, projects, and practices to ensure that differing needs, experiences, and perspectives of all genders are acknowledged and addressed. This happens, for example, by providing equal opportunities for education, employment, and leadership positions, and ensuring policies support work–life balance, such as parental leave for all genders. The future of CoPs after the RESET project largely depends on the organisations' commitment to fostering ongoing learning and collaboration. When well-supported, CoPs can evolve into powerful drivers of innovation, knowledge retention, and continuous improvement.

## Notes

1 https://act-on-gender.eu/communities-practice.
2 https://inspirequality.eu/.
3 https://gendersafe.eu/the-gendersafe-pledge-for-zero-tolerance-to-gender-based-violence.

## References

Brown, John, & Duguid, Paul. (1991). Organisational learning and communities of practice: Toward a unified view of working, learning, and innovation. *Organisation Science*, *2*(1), 40–57.

Davenport, Thomas, De Long, David, & Beers, Michael. (1998). Successful knowledge management projects. *Sloan Management Review Winter*, 43–57.

Dougherty, Deborah. (1995). Managing your core competencies for corporate venturing. *Entrepreneurship Theory and Practice*, *19*(3), 113–135.

Lämsä, Tuija. (2008). *Knowledge creation and organisational learning in communities of practice: An empirical analysis of a healthcare organisation*. Oulu: Oulu University Press.

Lämsä, Tuija. (2014). Managing knowledge and learning through communities of practice. In J. Alegre, R. Chiva, A. Fernández-Mesa, & J. L. Ferreras-Méndez (Eds.), *Shedding new lights on organisational learning, knowledge and capabilities*, 67–86. Cambridge Scholars Publishing.

Lämsä, Tuija, & Nätti, Satu. (2011). Knowledge sharing within and between communities of practice in a knowledge intensive organisation. In O. R. Hernáez & E. B. Campos (Eds.), *Handbook of research on communities of practice for organisational management and networking: Methodologies for competitive advantage*, 363–380. IGI Global.

Lesser, Eric, Fontaine, Michael, & Slusher, Jason. (Eds.). (2000). *Knowledge and communities*. Woburn: Butterworth-Heinemann.
Moran, John W., & Brightman, Baird K. (2001). Leading organizational change. *Career Development International*, *6*(2), 111–118.
Prahalad, C. K., & Hamel, Gary. (1990, May-June). The core competence of the corporation. *Harvard Business Review*, *68*, 79–91.
Snyder, William. (1996). *Organization, learning and performance: An exploration of the linkages between organization learning, knowledge, and performance*. Doctoral dissertation. Los Angeles: University of Southern California.
Wenger, Etienne. (1998, June). Communities of practice: Learning as a social system. *Systems Thinker*.
Wenger, Etienne. (2000). Communities of practice and social learning systems. *Organisation*, *7*(2), 225–246.
Wenger-Trayner, Etienne, & Wenger-Trayner, Beverly. (2015). *Introduction to communities of practice – a brief overview of the concept and its uses*. https://www.wenger-trayner.com/introduction-to-communities-of-practice

# INDEX

Note: Page locators in *italic* indicate a figure, and page locators in **bold** indicate a table on the corresponding page.

For Product Safety Concerns and Information please contact our EU representative GPSR@taylorandfrancis.com Taylor & Francis Verlag GmbH, Kaufingerstraße 24, 80331 München, Germany